ROBOTICS RESEARCH AND TECHNOLOGY

ROBOTICS

NEW RESEARCH

ROBOTICS RESEARCH AND TECHNOLOGY

Additional books in this series can be found on Nova's website under the Series tab.

Additional e-books in this series can be found on Nova's website under the e-book tab.

ROBOTICS RESEARCH AND TECHNOLOGY

ROBOTICS

NEW RESEARCH

DANIEL G. FISHER
EDITOR

nova publishers
New York

Additional color graphics may be available in the e-book version of this book.

Library of Congress Cataloging-in-Publication Data

Names: Fisher, Daniel G., editor.
Title: Robotics : new research / [editor] Daniel G. Fisher.
Other titles: Robotics (Nova Science Publishers)
Description: Hauppauge, New York : Nova Science Publisher's, Inc., [2016] | Series: Robotics research and technology | Includes bibliographical references and index.
Identifiers: LCCN 2016035633 (print) | LCCN 2016047109 (ebook) | ISBN 9781634859677 (hardcover) | ISBN 9781634859868
Subjects: LCSH: Robotics.
Classification: LCC TJ211 .R5679 2016 (print) | LCC TJ211 (ebook) | DDC 629.8/92--dc23
LC record available at https://lccn.loc.gov/2016035633

Published by Nova Science Publishers, Inc. † New York

CONTENTS

PREFACE

This book provides current research on robotics. The first chapter focuses on robot visual perception for object detection and recognition. Chapter Two discusses recent progress in vision-based robotic person following. Chapter Three examines portrait vision fusion. Chapter Four presents a preliminary design and experimental results of sex recognition. Chapter Five introduces the significance and application of health monitoring methods, and some of the traditional health monitoring methods are introduced. Chapter Six presents a systematic review of upper arm exoskeletons to find out under what clinical conditions use of such devices may be beneficial, what could be the technical requirements and what user interface must be provided to enhance their acceptability.

Chapter 1 - Today's robots have found their way from sealed working stations in factories to people's living and working spaces, where they should be able to autonomously perform different services useful to the well-being of humans such as domestic tasks, healthcare services, entertainment, and education. So, research is going from special-purpose service robots such as autonomous cleaning or transport systems, to multi-functional robot systems. In this context, a key issue is the interaction with objects of varying size, shape and degree of mobility. For that, autonomous service robots must be provided with the ability to properly process sensory data in real-time such that they can adequately react and quickly adapt to changes in the environment. In particular, this chapter will be focused on visual perception for object detection and recognition since there are still some open issues to be addressed when dynamic, unpredictable and human-oriented scenarios are considered. Thus, this chapter will review the existing state-of-the-art approaches when different sensory data is used, while analysing their behaviour in the face of different conditions and situations.

Chapter 2 - The chapter discusses recent progress in vision-based robotic person following. After briefly explaining the developments in the past decade, several approaches to person detection and following are described. Color based person detection in image using methods of color tolerance, color ellipsoid and color histogram are described. The effects of light changes on color-based person detection are explained and a method to compensate for significant environment light changes on the person's image is proposed. The camera internal gain and shutter speed controls enable person following when the person move from bright (sunny) areas to darker locations (shade) and vice versa. Region shape characteristics provide complementary information to color criteria for person identification, and a number of shape measures are suggested. In addition, a technique based on edge detection and Hough transform is described that allows color-independent person detection to complements the color dependent detection. In order to reduce computational burden and achieve real-time person following, a method is described to limit the area in the image that is searched to locate the person. This area is expanded or contracted adaptively based on detection confidence of the previous frames. In order to follow the person while keeping a reasonably constant distance from it, the robot speed and steering, and the camera pan and tilt are controlled. The control loops use information about the area and centroid of the detected upper body, e.g., torso or head in the image. Experiments are conducted using two robotic platforms; a Segway RMP and an ActiveMedia All-terrain rover. Experimental results are reported in various indoors and outdoors environments.

Chapter 3 - Video communications is a common way to communicate by the interactive technology, specially uses webcam for the remote interaction and see each other's characteristics from the screen display. In this research, the main goal is to augment some dynamic interactive virtual environments under robot vision system. For this achievement, a method using superimposing a segmented human portrait on a panoramic background is proposed, then the limb interactive element is added into these videos involved with a dynamic portrait segmentation method meanwhile uses Kinect device to extract a portrait for amendment, finally acquires a full of portrait information. Because the face is the most important identification region in one portrait, then the head skeleton tracking method is also used to strengthen the repair for its head segmentation, further uses the edge transparent processing to synthesis them into the video, therefore users can verbally and physically communicate through these video interactive ways to show more much vibrant.

Chapter 4 - Detecting human sex from complex background, illumination variations and objects under robot vision system is very difficult but important for an adaptive information service. In this research, a preliminary design and experimental results of sex recognition will presented from the walking movements that utilize the gait-energy image (GEI) with denoised energy image (DEI) pre-processing as a support vector machine (SVM) classifier to train and extract its characteristics. The result shows that the proposed method will adopt some characteristic value and the accuracy can reach up to 100% of sex recognition rate when people are walking at a fixed angle. Meanwhile, it will be able to achieve over 80% rate within some allowed fault-tolerant angle range.

Chapter 5 - Currently, the security of large civil engineering structures more and more people's attention, health monitoring study civil engineering at home and abroad is one of the hot topics of research. After a hundred years of research, the traditional structural health monitoring theories and methods continue to improve, and now there have been a lot to be able to use the results of the actual engineering and have been widely applied to large bridges and other structures, but there is a lot of its own defect. With the development of computer technology, people began to use computer vision to health monitoring, and gradually formed a monitoring method of mechanical vision. This paper introduces the significance and application health monitoring methods, and some of the traditional health monitoring methods are introduced. Then reviewed the development and the use of machine vision health status monitoring, and finally to crack measuring, for example, using wavelet recognition method for image recognition and analysis of fracture.

Chapter 6 - There have been many upper arm exoskeletons developed in the last two decades however none of these have met the expectations of the real users. Most of these designs did not go beyond the laboratory environment and were mainly used either as a research tool or as experimental devices for a very specific application. The majority of these devices were platform-based that was one of the main inhibiting factors why users found such devices too cumbersome to use. Even some of the portable arm exoskeletons did not find much favourable acceptance amongst the user groups because they were too heavy or difficult to put on and operate. This opens up a question, what specifications of upper arm exoskeletons will meet users' acceptability? This article presents a systematic review of upper arm exoskeletons to find out under what clinical conditions use of such devices may be beneficial, what could be the technical requirements and what user interface must be provided to enhance their acceptability. This paper looks into all major aspects of design related issues of upper arm exoskeleton that existed and the challenges ahead to overcome from technical as well as clinical application point of view. Whether arm exoskeletons can be used for stroke patients, Parkinson's or patients with weak musculature in assistive mode or for rehabilitation training? Also for those patients who suffer from different neuromuscular disorders have stiff muscles resulting in arm spasticity that doesn't allow free movement of the joint therefore the exoskeleton design should be adaptable to such clinical needs as well. The physiological parameter should be set as a standard for measuring patient status as the rehabilitation progresses and for bio-feedback. On the technological aspects the design should considers arm's degrees of freedom, dependency of joints and control regime to justify the optimal design performance. It should be a portable system with active and passive joints with gravity compensation provision to have low energy consumption. The type of actuator selection is one of the important criteria for the acceptability of such devices. It must also consider reconfiguration of the exoskeleton for misalignment, singularity, redundancy in design and weight. There are a large number of sophisticated control algorithms existing but most of them are not adaptive to safety measures and for the ease of operation. In addition to the above, the design must also consider vitally important user interface particularly with the advent of inexpensive electronics gadgets such as Mayo Armband, Google glasses, Eye Tracking systems and Kinect for motivating user rehabilitation.

In: Robotics: New Research
Editor: Daniel G. Fisher

ISBN: 978-1-63485-967-7

Chapter 1

OBJECT RECOGNITION FOR ROBOT TASKS: AN OVERVIEW

Ester Martinez-Martin*[*] *and Angel P. del Pobil
Robotic Intelligence Lab (RobInLab), Universitat Jaume-I,
Castellón, Spain

ABSTRACT

Today's robots have found their way from sealed working stations in factories to people's living and working spaces, where they should be able to autonomously perform different services useful to the well-being of humans such as domestic tasks, healthcare services, entertainment, and education. So, research is going from special-purpose service robots such as autonomous cleaning or transport systems, to multi-functional robot systems. In this context, a key issue is the interaction with objects of varying size, shape and degree of mobility. For that, autonomous service robots must be provided with the ability to properly process sensory data in real-time such that they can adequately react and quickly adapt to changes in the environment. In particular, this chapter will be focused on visual perception for object detection and recognition since there are still some open issues to be addressed when dynamic, unpredictable and human-oriented scenarios are considered. Thus, this chapter will review

[*] Corresponding author: Ester Martinez-Martin. Email: emartine@uji.es

the existing state-of-the-art approaches when different sensory data is used, while analysing their behaviour in the face of different conditions and situations.

Keywords: object recognition, robotics, survey

INTRODUCTION

Autonomous robot systems should be endowed with the ability to move through an environment while interacting with it. This interaction implies to be aware of what is around them, where the target objects are located, where they can move and when they must adapt to changes in the environment. In this context, perception plays a main role, becoming a basic requirement for any robotic system working in real scenarios.

Generally speaking, perception can be understood as the processing of the data from external sensors and their interpretation. Clearly, the perception process takes different forms depending on the required output. In particular, given its great importance in the Robotics research area, this chapter is focused on object recognition, understood as 'the ability to distinguish the occurrence of a stimulus that was previously presented from one that was not' [1]. In this sense, although an object can be recognised at different levels, the two most commonly-used are (see Figure 1):

- *Instance level*. This is the specific case where the goal is to identify if an object is physically the same object that was previously observed (i.e., instances of the same object). So, the recognition level corresponds to answering questions like *is this Angel's coffee mug or Ester's?*
- *Category level*. This recognition level involves identifying different instances of a generic category as belonging to the same conceptual class. Thus, the question to be answered takes the form *is this an orange or a glass?*

Although human beings recognise objects effortlessly even when the objects vary somewhat in form, colour, texture, size, pose, etc., the algorithmic description of this task has been very difficult, especially when the object's appearance changes. Despite the wide research on this topic, there are still some challenges to be overcome. This chapter is aimed to give an overview of the state-of-the-art approaches when different sensory data is used. So, firstly, visual processing for object recognition is analysed. Then, the object recognition is studied from the tactile point of view. Later, multisensory object recognition approaches are discussed. Finally, conclusions are presented.

Figure 1. The two most common object recognition levels: (1) identifying instances of a particular object (top); and, (2) categorising instances of the same object class (bottom).

OBJECT RECOGNITION USING VISUAL INFORMATION

Traditionally, vision has been used for object recognition due to the rich information it can provide. However, the availability of a considerable amount of information about the environment, occlusions, deformations, and changes in image contrast, intensity, shadow, colour, scale, viewpoint (e.g., rotation, translation, etc.), make the visual perception a complex problem. Despite the wide research in this area, the existing literature for robot's visual perception can be grouped in three major categories:

1. Visual perception based on object model
2. Visual perception based on object features
3. Human-like perception based on selective attention mechanisms

Model-based techniques consist of recognising objects by matching an object model with the local model obtained from the visual data. If the model is not known, a learning step is performed in advance so that an object model is generated from a large set of training images showing the object from different viewpoints and in different poses. So, a correspondence between the model and the object features is used to recognise the interest object and its pose. Note that the feature selection plays a main role in object recognition since they determine the success of the process. As an example, Giessen and Schmidhuber [2] presented a technique for estimating and recognising the 3D object shape from colour features. Despite the great information provided by the colour cue, textureless objects and those without colour discriminative enough from the background, can make the recognition process fail.

Other approaches learn the 3D object shape from a sequence of 2D images. Although it properly works for some kind of objects, the high dimensionality of an object or its rich variability in its motion can result in wrong object learning. As a solution, a simplified model, such as points, stick, ellipsoidal, cylindrical, skeleton or shape-from-silhouettes, can be used [3-11]. Even though good results have been obtained in laboratory settings with static cameras, they do not usually operate in real-time and often rely on a static, uniform background. Moreover, it is not applicable to the real world since target models must be known a priori.

Some additional challenges arise when flexible objects must be handled in a dynamic, unconstrained situation. For instance, material deformation must be modelled and sensed for an accurate object manipulation. However, that analytic modelling could vary along time and, therefore, some kind of adaptability must be integrated in the robotic system. This is the underlying idea of the work presented by Bodenhagen et al. [12] to perform grasping and two manipulation tasks (a Peg-in-Hole insertion action and a Laying-Down operation) with flexible objects in a robotics workcell. So, the process is composed of four stages. Firstly, a 3D object model is generated from an Kinect RGB-D depth image. Then, the model deformation is tracked by using non-uniform rational B-splines (NURBS) function in the generic case [13] or a simple bending deformation in the special case to track the object deformation when a robot gripper on one end grasps it [14]. In a second stage, a subjacent deformation model is fitted to the observed data from the visual deformation along with the force reading by using an indenter. Thirdly, a complete object spatio-temporal description of the elastic state is generated by means of the general elasticity equations. Finally, a learning schema based on the action to be performed, the object itself and the current state of the object, is applied to properly manipulate an object.

Giving a next autonomy level, feature-based methods represent an object based on visual features such as colour, edges, corners, etc. That is, the objects of interest are represented by abstract characterizations. Such characterizations are essentially defined by measurable visual features and its similarity with the input data allows the robot system to identify the observed object. Consequently, the ability to uniquely represent an object from the visual input determines the technique's efficiency for object recognition. In this regard, several abstraction levels have been explored. So, at first level, features such as colour or location are obtained from the pixel information. So, colour can be a distinctive object feature for object recognition. Actually, it is one of the early, commonly-used object recognition techniques in the literature. However, colour information is very sensitive to photometric variations and changes in illumination and geometry. An RGB normalization could be used to partially achieve the illumination variance [15, 16], although a high invariance level can be obtained from a different colour representation [17, 18]. In any way, texture information is not captured by colour information at pixel level. That is, only objects homogeneously coloured and distinctive enough from the background can be recognised with this kind of approaches.

Other visual features can be considered. For example, motion has been considered as primary cue because it allows a robotic system to obtain other charactersitics such as object shape, speed or trajectory, which are meaningful for detection and recognition. Note that the motion observable in a visual input could be due to different factors: movement of the imaged objects (targets and/or vacillating background elements), movement of the observer, motion of the light sources or a combination of (some of) them. Therefore, image analysis for motion detection will be conditional upon the considered factors [19, 20]. Nevertheless, the interest objects are not always in movement, what makes these techniques useless for some robotics tasks like robot manipulation.

As an alternative, a large variety of object descriptors have been proposed in the literature such as edges [21], contours [22], shapes [23], Gaussian derivatives [24], complex features [25, 26], steerable filters [27, 28], or phase-based local features [29]. Primary cue combination can be also possible to provide more robustness to the object recognition process [30, 31, 32]. Nevertheless, one of the most well-known is the Scale Invariant Feature Transform (SIFT) approach introduced by Lowe [33, 34] which shares many features with neuron responses in primate vision. Basically, this descriptor, which has been shown to outperform others as [35], is a 4-stage filtering approach that detects the dominant gradient orientation at its location and provides a feature description of an object. That feature array allows a system to locate the target object in an image containing many other objects.

However, it imposes a large computational load, especially for real-time systems, or for low-power devices such as cellphones. Various refinements on this basic scheme have been proposed. Ke and Sukthankar [36] applied PCA on the gradient image around the detected interest point. This PCA-SIFT yields a 36-dimensional descriptor which is fast for matching (although is slower in feature computation), but proved to be less distinctive than SIFT in a comparative study [37]. In that comparative study [37], Mikolajczyk proposed another variant of SIFT, called GLOH, which proved to be even more distinctive with the same number of dimensions. However, GLOH is computationally more expensive as it uses again PCA for data compression. Se et al. [38] implemented SIFT on a Field Programmable Gate Array (FPGA) and improved its speed by an order of magnitude. Meanwhile, Grabner et al. [39] also used integral images to approximate SIFT. Their detection step is based on difference-of-mean (without interpolation) and their description step on integral histograms. Although it results in a faster method, it is at the cost of reduced quality compared to SIFT. On their behalf, Abdel-Hakim and Farag [40] added colour information to the SIFT descriptor making it more robust to colour and photometric variations, but still suffers the remaining SIFT problems.

Later, Bay et al. [41], focused on scale and in-plane rotation invariant detectors and descriptors, developed a scale- and rotation-invariant detector and descriptor, coined Speeded-Up Robust Features (SURF). Mainly, an input image is analysed at different scales in order to guarantee invariance to scale changes. Then, the detected interest points are provided with a rotation and scale invariant descriptor to finally apply a simple and efficient first-line indexing technique, based on the contrast of the interest point with its surrounding. Although computational time has been drastrically reduced, most notably with GPU devices [42], it still works on feature arrays to describe the objects in a scene, and the issue to properly identify those features in any scene has not been solved. So, despite a good performance was obtained in real textured and structured scenes, it has problems to extract features on textureless scenes or on low-resolution images.

One way to accelerate the matching process and reduce the memory consumption is to work with short descriptors. This is the case of Calonder et al. [43], who presented a binary descriptor relying on a relatively small number of intensity difference tests to represent an image patch as a binary string. Despite the good performance of the Binary Robust Independent Elementary Features (BRIEF), it is dependent on the image scale and rotation. As an improvement, Rublee et al. [44] presented a keypoint descriptor based on the FAST keypoint detector [45] and the BRIEF descriptor [43]. So, the implemented Oriented FAST and Rotated BRIEF (ORB) uses intensity centroid and image pyramids to provide rotation and scale invariance, considerably improving the performance of its component techniques and its efficiency, as demonstrated in [46]. As in the case of SIFT, the number of keypoints to be used is variable and must be set for each interest object, what could make the system fail if it is too low.

Local feature extraction and comparison can be also applied to 3D range data. One of the most popular descriptors is matching object surfaces by matching their points using their corresponding spin image[1] representation [47]. In spite of its good performance, there are some issues to be addressed such as the parametrizations, the automated learning of descriptive spin images, the use of multiresolution spin images, or the treatment of the empty space (e.g., beyond object borders).

On the way to more general object recognition, Gelfand et al. [48] presented a global registration using Integral Volume Descriptors (IVD) estimated at certain interest points in the data. Those interest points are extracted using a self similarity approach in the IVD space, meaning that the point descriptor is compared to the descriptors of its neighbours to determine areas where there is a significant change. While this method for interest point extraction explicitly takes the descriptor into account, it becomes impractical for more complex descriptors, which are more expensive to extract. As an improvement, Steder et al. [49] proposed Normal Aligned Radial Feature (NARF), a method for interest point extraction combined with a feature descriptor for points in 3D range data. For that, this tecnique searches points that are in the vicinity of significant changes and at the same time are on stable parts of the object surface. In addition, border information is explicitly used. At this point, it is important to take into account that, although depth information can be less sensitive to lighting conditions and scale information is directly available, object geometry is less expressive in terms of uniqueness. In addition, the presence of noise can considerably reduce the performance in a recognition process, as Rangel et al. broadly discussed in [50]. Their study highlights that the use of the Growing Neural Gas (GNG) method improves recognition rates due to the noise reduction without significant information lost, obtaining even better results than those of Voxel grid and/or of the raw clouds.

---

[1] A spin image is a data level shape descriptor that is used to match surfaces represented as surface meshes.

On the other hand, visual global features refers to observe the input image as a whole and extract a profile condensing the image's statistics. In this way, an entire object can be generalised with a single vector (e.g., moment invariants [51]), making the standard classification easier, except for the case a variable number of features is available per image. Therefore, techniques based on visual global features are useful in applications where it is possible to get a rough segmentation of the interest object. Thus, for instance, template matching could be used since it searches small image parts that match with any of the stored templates through similarity measurements such as Sum of Squared Differences (SSD) or Normalized Cross Correlation (NCC). Note that this kind of approaches are successful when the normal template is used. Nonetheless, they fail when some altered template image is used (e.g., different size, rotation, etc.) as shown in [52]. Other widely used feature extraction approaches include colour histograms [53], gradient histograms [54], objectness [55-58], spherical harmonic invariants [59], Fast Point Feature Histogram (FPFH) [60], and Viewpoint Feature Histogram (VFH) [61]. This kind of approaches has been also applied to other robot applications such as robot localization [62] and navigation [63]. A disadvantage of using global features is that the local environment information is lost such that it is difficult to produce precise robot actions. In addition, this kind of visual features are sensitive to clutter and occlusion.

Given that the pixel representation of an object might dramatically change when the object's pose varies, some improvements have been proposed. For example, hybrid object descriptors leveraging the benefits of all the individual descriptors, can also be defined. Several examples can be found in the literature: colour and edge [64]; colour and contours [65]; local and global edges [66]; local and global features [67, 68]. In addition, visual features dependent on a specific domain such as face [69] or fingerprint recognition, could be considered. Another solution could be to use images taken with the target object in several poses for training. Nevertheless, it seems compulsory to separate the invariance learning from the object recognition problem in order to be able to achieve a human-like recognition capability. As an early step, convolutional neural networks were used for dealing with shift invariance so that invariance was learnt as part of classification [70-72]. Alternatively, an invariant representation of the environment can be obtained from temporal input sequences when it is assumed that code for local properties varies quickly whereas the perceived environment changes slowly [73]. Nevertheless, as demonstrated by [74], their application is very limited when they learn global invariance because of the adaptation to a single transformation group.

Going a step further, Zou et al. [75] presented a hierarchical, modular network to capture higher levels of invariance (e.g., out-of-plane transformations or non-linear warping) thanks to the learning of multiple layers of representations.

From research in Psychology and Physiology, human-like visual perception can be developed for robots by a visual attention mechanism which acts as a bridge betweeen perception and action. So, this selective attention mechanism is used to reduce the image processing to those objects that are relevant to the present task and situation [110, 111, 84]. In this way, a robot system might be endowed with the cognitive ability to autonomously learn and reason about how to perceive its surrounding environment. So, the focus of attention (FOA) in humans is determined through two ways: the bottom-up attention (an automatic behaviour directing the FOA based on object's conspicuousness in temporal and spatial context), and the top-down attention (a goal-directed behaviour guiding the FOA according to the knowledge, current tasks and context of the mental states). Furthermore, the attentional selection is either space-based or object-based. The fundamental difference between them is the underlying unit of attentional selection. So, the space-based attention theory establishes that attention is directed to a spatial location [76-78]. On the contrary, object-based attention theory states that some pre-attentive processes are used to segment the visual field into discrete proto-objects[2], followed by attention that deals with one proto-object at a time [80–84]. That is, the attentional selection and the perceptual organization must go together [85].

From these statements, some research has been carried out. Itti et al. [86] built a space-based bottom-up attention model based on feature integration theory (FIT) [87]. With the purpose of modelling the bottom-up attention in terms of both spatial and temporal context, the surprise mechanism was proposed [88-90]. Based on the guided search model (GSM) [91], Navalpakkam and Itti [92] extended the Itti's model by modelling the top-down attention. Tsotsos et al. [93] presented a model that selectively tunes the visual processing networks by a top-down hierarchy of winner-take-all processes. Later, Chikkerur et al. [94] proposed a computational method by modelling attention as a Bayesian inference process. Alternatively, Sun and Fisher [99] proposed a computational method for object-based bottom-up

[2] A proto-object can be defined as a homogeneous region obtained by Gestalt principle [79], e.g., proximity and similarity rules.

attention. In the Sun's model, the saliency of a proto-object is estimated by combining the space-based saliency of that proto-object.

Some visual attention models were also proposed for robots by integrating both bottom-up and top-down attention [95-98], while others have been applied those attention models in object detection [100, 101], object recognition [102], object tracking [103, 104], exploration of unknown environment [105] or virtual environment [106], attentional search [107], and visual perception of robots [108, 109].

Another alternative for object recognition is to label the interest objects with an identification tag so that the robotic system visualises and interprets the corresponding label. This is, for instance, the case of LIBO [112], a mobile robotic system for a library that uses an android phone camera to read the book barcodes, compares it with the required one and generates the signal on arduino microcontroller for the 'pick-and-place' action. Sathya et al. [113] used an RFID tag to identify missplaced and missing books in a library. On their behalf, Prats et al. [114] presented a complete robotic system to assist users in the library at Universitat Jaume I (UJI). In this case, each book in the library contains a tag in its spine with its corresponding signature according to the US Library of Congress Classification (LCC) so that a probabilistic tracking method to match labels through time, exploiting the spatio-temporal constraints of the problem, is used. Basically, book labels are segmented from the taken images and sent to an OCR module for code identification. Even though this is a robust way to recognise objects, it considerably restricts the robot system to only interact with the tagged objects, making it inappropriate for real applications.

Object Recognition Using Tactile Information

Human beings are excellent at recognising real-world objects from its stiffness and weight, on how it deforms on contact, how its surface feels when touched, and how it moves when pushed, by using the touch sensing alone [115, 116]. In the context of robotics tasks, tactile sensing has been used in different tasks such as exploring 3D object shapes [117], detecting ridges and bumps in the material by sliding a robotic finger over the object [118], determining the hardness/softness of probed objects [119], controlling the robot's position [120], or estimating the 3D pose of objects with known shapes [121].

From an object recognition point of view, Russel [122, 123] proposed an approach that extracted geometric features such as point, line or area contacts, and integrated them to classify objects into generic classes (e.g., boxes, spheres, cylinders, etc.). Later, Schneider et al. [124] presented an approach for object recognition using tactile observations from the touch-sensitive fingers of a manipulation robot. Thus, given the low-intensity images resulting from the touch sensory data, a k-means is applied as unsupervised clustering on the training dataset to build a vocabulary for the bag-of-features classifier. From that vocabulary, a codebook histogram can be learnt and used for object recognition, requiring a small number of grasp actions performed at different positions.

A different approach was presented by Bierbaum et al. [125]. In this case, a tactile exploration strategy based on dynamic fields for motion guidance of the robotic hand, is used to guide an antropomorphic five-finger hand along the surface of (unknown) objects with the aim of building a 3D object representation based on the acquired tactile point clouds. On the contrary, Gorges et al. [126] used haptic key primitives as a compact object description so that an object can be recognised from the finger positions and tactile patterns of the anthropomorphic robot hand.

MULTISENSORY OBJECT RECOGNITION

Even though some objects can be clearly recognised by using one of the human-like senses, the combination of (some of) them considerably enrich the object recognition. For instance, there are many common objects visually similar, but easily distinguishable from a tactile point of view such as ripe versus unripe fruits. So, these two sensory data could be merged to obtain a more robust object recognition. This is the case of Liu et al. [127] who, under the assumption that visual and tactile information is required for a proper object recognition, have developed a fusion framework for industrial manipulation scenarios in structured environments. In rough outlines, this train-test approach starts when one image sample (a covariance descriptor generated from the visual input) and one tactile sample (a multivariate-time-series model representing the tactile data) are taken. As those samples are usually gathered separetely, a Joint Group Kernel Sparse Coding (JGKSC) method is used to solve the intrinsically weak pairing. Finally, a classifier obtained from the previously acquired training samples, recognises an object.

Alternatively, other sensory data could be integrated. For instance, thermal imagery provides complimentary information to RGB data, especially under low visibility conditions. This is the criteria used by Kus [128] for pattern recognition in adverse weather conditions. In this case, object features are described by pattern vectors containing moment invariants, separately obtained from the visual and thermal images. Then, the distance between the stored pattern vectors and the current ones in each kind of image, is estimated and used for object recognition.

Conclusion

Object recognition is of great importance for any robotic task since it is the way any robot interacts with its environment. However, although human beings effortlessly recognises any daily object, there are several factors that might make the robot system miserably fails such as object intra-class similarities or changes in the appearance, viewpoint, scale, pose, illumination conditions, etc. Motivated by these issues, scientific community deeply investigates this topic, addressing it from different point of views and disciplines.

A sample of that is presented in this chapter, where different sensory data processing as well as their combination has been examined. In particular, this chapter has been focused on vision, which has extensively studied the object recognition problem, touch, that recently played a major role on this area, and the combination of them, enriching the sensory data for a proper object recognition. Great advances have been done in all these lines. However, there are still far from solving all the presented challenges when real scenearios are considered.

References

[1] Squire, L.R., Wixted, J.T., Clark, R.E., 2007). Recognition memory and the medial temporal lobe: A new perspective. *Nature Reviews Neuroscience*, 8(11), 872–883.

[2] Giessen, M., Schmidhuber, J., 2005. *Fast Color-Based Object Recognition Independent of Position and Orientation*, Intl Conf on Artificial Neural Networks: Biological Inspiration (ICANN), 469–474.

[3] Truong, H.Q., Lee, S., Jang, S.-W., 2008. *Model-based Recognition of 3D Objects using Intersection Lines*, IEEE Intl Conf on Multisensor Fusion and Integration for Intelligent Systems, 656–660.

[4] Kamgar-Parsi, B., Kamgar-Parsi, B., May 2004. Algorithm for matching 3D line sets, *IEEE Trans on Pattern Analysis and Machine Intelligent*, 26(5), 582–593.

[5] Weiss, I., Ray, M., 2001. Model-based Recognition of 3D Objects from Single Images, *IEEE Trans on Pattern Analysis and Machine Intelligence (PAMI)*, 23(2), 116–128.

[6] Breglera, C., Malik, J., Pullen, K. 2004. Twist based acquisition and tracking of animal and human kinematics, *Intl. Journal of Computer Vision (IJCV)*, 56(3), 179–194.

[7] Urtasun, R., Fleet, D.J., Fua, P. 2006. Temporal motion models for monocular and multiview 3D human body tracking, *Computer Vision and Image Understanding (CVIU)*, Special issue on modeling people: Vision-based understanding of a person's shape, appearance, movement and behaviour, 104(2), 157–177.

[8] Mikic, I., Trivedi, M., Hunter, E., Cosman, P. 2003. Human body model acquisition and tracking using voxel data, *Intl. Journal of Computer Vision (IJCV)*, 53(3), 199–223.

[9] Bobick, A.F., Davis, J.W. 2001. The Recognition of Human Movement Using Temporal Templates, *IEEE Trans on Pattern Analysis and Machine Intelligence (PAMI)*, 23(3), 257–267.

[10] Cheung, K., Baker, S., Kanade, T., 2005. Shape-from-silhouette across time part I: Theory and Algorithms, *Intl. Journal of Computer Vision (IJCV)*, 62(3), 221–247.

[11] Cristinacce, D., Cootes, T.F., Sep 2007. *Boosted Regression Active Shape Models*, British Machine Vision Conference, vol. 2, 880–889.

[12] Bodenhagen, L., Fugl, A.R., Jordt, A., Willatzen, M., Andersen, K.A., Olsen, M.M., Koch, R., Petersen, H.G., Krüger, N., Jul. 2014. An adaptable robot vision system performing manipulation actions with flexible objects. *IEEE T. Autom. Sci. Eng.*, 11 (3), 749–765.

[13] Jordt, A., Koch, R., Sep. 2011. *Fast tracking of deformable objects in depth and color video*. British Machine Vision Conf., BMVC'11, 114.1–114.11.

[14] Fugl, A.R., Jordt, A., Petersen, H.G., Willatzen, M., Koch, R., Aug. 2012. *Estimation of material properties and pose for deformable objects from depth and color images*. Joint Pattern Recognit. Symp. (DAGM-OAGM), Graz, Austria, 165–174.

[15] Funt, B.V., Barnard, K., Martin, L., 1998. *Is machine colour constancy good enough?*, European Conference on Computer Vision (ECCV), 445–459.

[16] Drew, M.S., Wei, J., Li, Z-N., 1998. *Illumination-invariant color object recognition via compressed chromaticity histograms of color-channel-normalized images*, Intl Conf on Computer Vision (ICCV), 533–540.

[17] Geusebroek, J.M., van den Boomgaard, R., Smeulders, A.W.M., Geerts, H., Dec 2001. Color invariance, *IEEE Trans. Pattern Anal. Machine Intell. (PAMI)*, 23(12), 1338–1350.

[18] Gevers, T., Smeulders, A.W.M., 1999. Color-based object recognition, *Pattern Recognition*, 32(1999), 453–464.

[19] Ryu, S.P., Kwak, N.J., 2015. Camera Control Method for Object Recognition System, *Information*, 18(10), 4281–4287.

[20] Martinez-Martin, E., del Pobil, A.P., 2012. Robust Motion Detection in Real-Life Scenarios, Springer.

[21] Shukla, A.J., Patel, V., Gajjar, N., 2015. *Implementation of Edge Detection Algorithms in Real Time on FPGA*, 5th Nirma University International Conference on Engineering (NUiCONE).

[22] Stiene, S., Lingemann, K., Nuchter, A., Hertzberg, J., 2006. *Contour-Based Object Detection in Range Images*, Intl Symp on 3D Data Processing, Visualization, and Transmission, 168–175.

[23] Belongie, S., Malik, J., Puzicha, J., Apr 2002. Shape Matching and Object Recognition Using Shape Contexts, *IEEE Trans on Pattern Analysis and Machine Intelligence (PAMI)*, 24(24), 509–522.

[24] Florack, L.M.J., ter Haar Romeny, B.M., Koenderink, J.J., Viergever, M.A. 1994. General intensity transformations and differential invariants. *Journal of Mathematical Imaging and Vision (JMIV)*, 4(2), 171–187.

[25] Schaffalitzky, F., Zisserman, A., 2002. *Multi-view matching for unordered image sets, or "How do I organize my holiday snaps?,"* European Conference on Computer Vision (ECCV), vol. 1., 414–431.

[26] Baumberg, A. 2000 (June). Reliable feature matching across widely separated views. *Pages 774–781 of: IEEE Conference on Computer Vision and Pattern Recognition (CVPR)*, vol. 1.

[27] Villamizar, M., Sanfeliu, A., Andrade-Cetto, J., Sep. 2006. *Computation of Rotation Local Invariant Features using the Integral Image for Real Time Object Detection.* 18th International Conference on Pattern Recognition (ICPR), vol. 4, 81–85.

[28] Freeman, W.T., Adelson, E.H., 1991. The design and use of steerable filters. *Pattern Analysis and Machine Intelligence (PAMI)*, 13(9), 891–906.

[29] Carneiro, G., Jepson, A.D., Jun 2003. *Multi-scale phase-based local features*. IEEE Conference on Computer Vision and Pattern Recognition (CVPR), vol. 1, 736–743.

[30] Lai, K., Bo, L., Ren, X., Fox, D., 2011. *Sparse Distance Learning for Object Recognition Combining RGB and Depth Information*, Intl Conf on Robotics and Automation (ICRA), 4007–4013.

[31] Diplaros, A., Gevers, T., Patras, I., Jan 2006. Combining Color and Shape Information for Illumination-Viewpoint Invariant Object Recognition, *IEEE Trans on Image Processing*, 5(1), 1–11.

[32] Cucchiara, R., Grana, C., Piccardi, M., Prati, A., Sirotti, S., 2001. *Improving Shadow Suppression in Moving Object Detection with HSV Color Information*, IEEE Intelligent Transportation Systems, 334–339.

[33] Lowe, D.G., Sep 1999. *Object recognition from local scale-invariant features*. International Conference on Computer Vision (ICCV), 1150–1157.

[34] Lowe, D.G. 2004. Distinctive image features from Scale-Invariant Keypoints. *IEEE I. Comp. Vis.*, 60(2), 91–110.

[35] Mikolajczyk, K., and Schmid, C., Jun 2003. *A performance evaluation of local descriptors*. IEEE Conference on Computer Vision and Pattern Recognition (CVPR), vol. 2, 257–263.

[36] Ke, Y., Sukthankar, R., 2004. *PCA-SIFT: A more distinctive representation for local image descriptors*, IEEE Conference on Computer Vision and Pattern Recognition (CVPR), vol. 2, 506–513.

[37] Mikolajczyk, K., Schmid, C., 2003. A performance evaluation of local descriptors, *Pattern Analysis and Machine Intelligence (PAMI)*, 27(10), 1615–1630.

[38] Se, S., Ng, H.K., Jasiobedzki, P., Moyung, T.J., 2004. *Vision based modeling and localization for planetary exploration rovers*, 55th International Astronautical Congress.

[39] Grabner, M., Grabner, H., Bischof, H., 2006. *Fast approximated SIFT*. 7th Asian Conference on Computer Vision (ACCV), 918–927.

[40] Abdel-Hakim, A.E., Farag, A.A., 2006. *CSIFT: A SIFT descriptor with Color Invariant Characteristics*, IEEE Computer Society Conf on Computer Vision and Pattern Recognition (CVPR).

[41] Bay, H., Ess, A., Tuytelaars, T., Gool, L. Van., 2008. SURF: Speeded Up Robust Features. *Comput. Vis. Image Und. (CVIU)*, 110(3), 346–359.

[42] Sinha, S.N., Frahm, J.M., Pollefeys, M., Genc, Y., 2006. *Gpu-based video feature tracking and matching*. Technical report, In: Workshop on Edge Computing Using New Commodity Architectures.

[43] Calonder, M., Lepetit, V., Strecha, C., Fua, P., 2010. *Brief: Binary robust independent elementary features*. European Conf on Computer Vision (ECCV).

[44] Rublee, E., Rabaud, V., Konolige, K., Bradski, G.R., 2011. *ORB: An efficient alternative to SIFT or SURF*. Intl Conf on Comp Vision (ICCV), 2564–2571.

[45] Rosten, E., Drummond, T., 2006. *Machine learning for highspeed corner detection*. European Conf on Computer Vision (ECCV).

[46] Kulkarni1, A.V., Jagtap, J.S., Harpale, V.K., 2013. Object recognition with ORB and its Implementation on FPGA, *Intl. J. of Advanced Computer Research*, 3(11), 164–169.

[47] Johnson, A.E., Hebert, M., 1999. Using spin images for efficient object recognition in cluttered 3d scenes. *IEEE Trans. Pattern Anal. Mach. Intell. (PAMI)*, 21(5), 433–449.

[48] Gelfand, N., Mitra, N.J., Guibas, L.J., Pottmann, H., 2005. *Robust global registration*. The third Eurographics symposium on Geometry processing.

[49] Steder, B., Rusu, R.B., Konolige, K., Burgard, W., 2011. *Point feature extraction on 3D range scans taking into account object boundaries*, IEEE Intl Conf on Robotics and Automation (ICRA), 2601–2608.

[50] Rangel, J.C., Morell, V., Cazorla, M., Orts-Escolano, S., García-Rodríguez, J., April 2016. Object recognition in noisy RGB-D data using GNG. *Pattern Anal Applic*, 1–16.

[51] Mindru, F., Tuytelaars, T., Gool, L. Van, Moons, T. 2004. Moment invariants for recognition under changing viewpoint and illumination. *Computer Vision and Image Understanding (CVIU)*, 94(1–3), 3–27.

[52] Hisham, M.B., Yaakob, S.N., Raof, R.A.A. Nazren, A.B.A., Wafi, N.M., 2015. *Template Matching Using Sum of Squared Difference and Normalized Cross Correlation*, IEEE Student Conf on Research and Development (SCOReD), 100–104.

[53] Mekhtiche, M.A., Bencherif, M.A., Algabri, M., Alsulaiman, M., Hedjar, R., Faisal, M., AlMutib, K., Nov 2015. Real Time Object Detection and Tracking over a Mobile Platform, *Indian Journal of Science and Technology*, 8(32), 1–7.

[54] Shehnaz, M., Naveen, N., 2015. *An Object Recognition Algorithm with Structure-Guided Saliency Detection and SVM Classifier*. IEEE

International Conference on Power, Instrumentation, Control and Computing (PICC).

[55] Cheng, M.-M., Zhang, Z., Lin, W.-Y., Torr, P., 2014. *BING: Binarized Normed Gradients for Objectness Estimation at 300fps*, IEEE Conf on Computer Vision and Pattern Recognition (CVPR), 3286–3293.

[56] Endres, I., Hoiem, D., 2013. Category-Independent Object Proposals with Diverse Ranking, *IEEE Trans on Pattern Analysis and Machine Intelligence (PAMI)*, 36(2), 222–234.

[57] Uijlings, J., van de Sande, K., Gevers, T., Smeulders, A., 2013. Selective search for object recognition, *Intl. Journal on Computer Vision*, 104(2), 154–171.

[58] Alexe, B., Deselaers, T., Ferrari, V., 2012. Measuring the objectness of image windows, *IEEE Trans on Pattern Analysis and Machine Intelligence (PAMI)*, 34(11), 2189–2202.

[59] Burel, G., Henocq, H., 1995. Three-dimensional invariants and their application to object recognition, *Signal Process.*, 45,(1), 1–22.

[60] Rusu, R.B., Blodow, N., Beetz, M., 2009. *Fast Point Feature Histograms (FPFH) for 3D Registration*, Intl Conf on Robotics and Automation (ICRA), 3212–3217.

[61] Rusu, R.B., Bradski, G., Thibaux, R., Hsu, J., 2010. *Fast 3D Recognition and Pose Using the Viewpoint Feature Histogram*, IEEE/RSJ Intl Conf on Intelligent Robots and Systems (IROS), 2155–2162.

[62] Ulrich, I., Nourbakhsh, I., 2000. *Appearance-based place recognition for topological localization*, IEEE Intl Conf on Robotics and Automation (ICRA), 1023–1029.

[63] Remazeilles, A., Chaumette, F., Apr. 2007. Image-based robot navigation from an image memory, *Robot. Auton. Syst.*, 55(4), 345–356.

[64] Savitha, D.K., Nema, M.K., 2015. *Novel Hybrid Image Content Descriptor*, IEEE Intl Conf on Computer Graphics, Vision and Information Security (CGVIS), 264–268.

[65] Silva, A.S., Severgnini, F.M.Q., Oliveira, M.L., Mendes, V.M.S., Peixoto, Z.M.A., Mar 2016. Object Tracking by Color and Active Models Segmetation, *IEEE Latin America Transactions*, 14(3), 1488–1493.

[66] Won, C.S., Park, D.K., Park, S.-J., Feb 2002. Efficient Use of MPEG-7 Edge Histogram Descriptor, *ETRI Journal*, 24(1), 23–30.

[67] Murphy, K., Torralba, A., Eaton, D., Freeman, W., 2006. Object Detection and Localization Using Local and Global Features, Toward Category-Level Object Recognition, 382–400.

[68] Lisin, D.A., Mattar, M.A., Blaschko, M.B., Learned-Miller, E.G., Benfield, M.C., 2005. Combining Local and Global Image Features for Object Class Recognition, IEEE Computer Society Conference on Computer Vision and Pattern Recognition (CVPR) - Workshops, 47–47.

[69] Turk, M., Pentland, A., 1991. Eigenfaces for Recognition, *Journal of Cognitive Neuroscience*, 3(1), 71–86.

[70] LeCun, Y., Bottou, L., Bengio, Y., 1998. Haner, P.: Gradient-based learning applied to document recognition. *Proc. of the IEEE*, 86(11), 2278–2324.

[71] Olshausen, B.A., Anderson, C.H., Van Essen, D.C., 1993. A neurobiological model of visual attention and invariant pattern recognition based on dynamic routing of information. *J. Neurosci.*, 13 (11), 4700–4719.

[72] Fukushima, K., 1980. Neocognitron: A self-organizing neural network model for a mechanism of pattern recognition unaffected by shift in position. *Biol. Cybern.*, 36, 193–202.

[73] Wiskott, L., Sejnowski, T.J., 2002. Slow feature analysis: Unsupervised learning of invariances. *Neural Comput.*, 14(4), 715–770.

[74] Hocke, J., Martinetz, T., 2015. Learning Transformation Invariance from Global to Local, Workshop New Challenges in Neural Computing, 16–24.

[75] Zou, W.Y., Zhu, S., Ng, A.Y., Yu, K., 2012. *Deep Learning of Invariant Features via Simulated Fixations in Video*. Adv Neur In 25 (NIPS), 3212–3220.

[76] Downing, C.J., 1988. Expectancy and visual-spatial attention: Effects on perceptual quality, *J. Experimental Psychology: Human Perception and Performance*, 14(2), 188–202.

[77] Eriksen, C.W., James, J.D.S., 1986. Visual attention within and around the field of focal attention: A zoom lens model, *Perception and Psychophysics*, 40(4), 225–240.

[78] Posner, M.I., 1980. Orienting of attention, *Quart. J. Experimental Psychology*, 32(1), 3–25.

[79] Palmer, S.E., 1999. *Vision Science: Photons to Phenomenology*, Cambridge, MA: MIT Press.

[80] Moore, C.M., Yantis, S., Vaughan, B., 1998. Object-based visual selection: Evidence from perceptual completion, *Psychological Sci.*, 9 (2), 104–110.

[81] Egly, R., Driver, J., Rafal, R., 1994. Shifting visual attention between objects and locations, *J. Experimental Psychology: General*, 123(2), 161–177.

[82] Tipper, S.P., Weaver, B., Jerreat, L., Burak, A., 1994. Object-based and environment-based inhibition of return of visual attention, *J. Experimental Psychology: Human Perception and Performance*, 20(3), 478–499.

[83] Kanwisher, N., Driver, J., 1992. Objects, attributes, and visual attention: Which, what, and where, *Current Directions in Psychological Science*, 1 (1), 26–31.

[84] Duncan, J., 1984. Selective attention and the organization of visual information, *J. Experimental Psychology: General*, 113(4), 501–517.

[85] Scholl, B.J., 2001. Objects and attention: The state of the art, *Cognition*, 80(1–2), 1–46.

[86] Itti, L., Koch, C., Niebur, E., Nov 1998. A model of saliency-based visual attention for rapid scene analysis, *IEEE Trans. on Pattern Analysis and Machine Intelligence (PAMI)*, 20(11), 1254–1259.

[87] Treisman, A.M., Gelade, G., 1980. A feature integration theory of attention, *Cognition Psychology*, 12(1–2), 507–545.

[88] Baldi, P., Itti, L., 2010. Of bits and wows: A Bayesian theory of surprise with applications to attention, *Neural Networks*, 23, 649–666.

[89] Maier, W., Steinbach, E., 2010. A probabilistic appearance representation and its application to surprise detection in cognitive robots, *IEEE Trans. Autonomous Mental Development*, 2(4), 267–281.

[90] Itti, L., Baldi, P., 2009. Bayesian surprise attracts human attention, *Visual Res.*, 49(10), 1295–1306.

[91] Wolfe, J.M., 1994. Guided search 2.0: A revised model of visual search, *Psychonomic Bull. Rev.*, 1(2), 202–238.

[92] Navalpakkam, V., Itti, L., 2005. Modeling the influence of task on attention, *Vision Res.*, 45(2), 205–231.

[93] Tsotsos, J.K., Culhane, S.M., Wai, W.Y.K., Lai, Y., Davis, N., Nuflo, F., 1995. Modeling visual attention via selective tuning, *Artif. Intell.*, 78, 507–545.

[94] Chikkerur, S., Serre, T., Tan, C., Poggio, T., 2010. What and where: A Bayesian inference theory of attention, *Visual Res.*, 50(22), 2233–2247.

[95] Begum, M., Karray, F., Mann, G.K.I., Gosine, R.G., Oct 2010. A probabilistic model of overt visual attention for cognitive robots, *IEEE Trans. Syst. Man, Cybern.–Part B: Cybern.*, 40(5), 1305–1318.

[96] Belardinelli, A., Pirri, F., 2006. A biologically plausible robot attention model, based on space and time, *Cognitive Process.*, 7(Supplement 5), 11–14.

[97] Belardinelli, A., Pirri, F., Carbone, A., 2006. *Robot task-driven attention*, Int. Symp. Practical Cognitive Agents Robots, 117–128.

[98] Frintrop, S., 2005. Vocus: A visual attention system for object detection and goal-directed search, Ph.D. dissertation, Univ. Bonn, Bonn, Germany.

[99] Hoya, T., Feb 2004. Notions of intuition and attention modeled by a hierarchically arranged generalized regression neural network, *IEEE Trans. Syst. Man, Cybern.—Part B: Cybern.*, 34(1), 200–209.

[100] Sun, Y., Fisher, R., 2003. Object-based visual attention for computer vision, *Artif. Intell.*, 146(1), 77–123.

[101] Yu, Y., Gu, J., Mann, G.K.I., Gosine, R.G., Apr 2013. Development and Evaluation of Object-Based Visual Attention for Automatic Perception of Robots, *IEEE Trans. on Automation Science and Engineering*, 10(2), 365–379.

[102] Cao, X., Lin, R., Yan, P., Li, X., 2012. Visual attention accelerated vehicle detection in low-altitude airborne video of urban environment, *IEEE Trans. Circuits Syst. Video Technol.*, 22(3), 366–378.

[103] Walther, D., Rutishauser, U., Koch, C., Perona, P., 2004. *On the usefulness of attention for object recognition*, Workshop on Attention and Performance in Computational Vision at Eur. Conf. Comput. Vision (ECCV), 96–103.

[104] Frintrop, S., Konigs, A., Hoeller, F., Schulz, D., 2010. A component-based approach to visual person tracking from a mobile platform, *Int. J. Social Robot.*, 2, 53–62.

[105] Yu, Y., Mann, G.K.I., Gosine, R.G., 2010. *Target tracking for moving robots using object-based visual attention*, IEEE Intl Conf. on Intelligent Robotics Systems (IROS), 2902–2907.

[106] Carbone, A., Finzi, A., Orlandini, A., 2008. Model-based control architecture for attentive robots in rescue scenarios, *Autonomous Robots*, 24(1), 87–120.

[107] Hillaire, S., Lecuyer, A., Regia-Corte, T., Cozot, R., Royan, J., Breton, G., 2012. Design and application of real-time visual attention model for

the exploration of 3D virtual environments, *IEEE Trans. Visualization Comput. Graphics*, 18(3), 356–368.

[108] Belardinelli, A., Pirri, F., Carbone, A., Apr 2007. Bottom-up gaze shifts and fixations learning by imitation, *IEEE Trans. Syst. Man, Cybern.—Part B: Cybern.*, 37(2), 256–271.

[109] Breazeal, C., Edsinger, A., Fitzpatrick, P., Scassellati, B., 2001. Active vision for sociable robots, *IEEE Trans. Syst. Man, Cybern.—Part A: Syst. Humans*, 31(5), 443–453.

[110] Backer, G., Mertsching, B., Bollmann, M., Dec. 2001. Data- and model-driven gaze control for an active-vision system, *IEEE Trans. on Pattern Analysis and Machine Intelligence (PAMI)*, 23(12), 1415–1429.

[111] Miller, E.K., 1999. Prefrontal cortex and the neural basis of executive functions, In: *Attention, Space and Action: Studies in Cognitive Neuroscience*, G. W. Humphreys, J. Duncan, and A. Treisman, Eds. Oxford, UK: Oxford Univ. Press, 251–272.

[112] Tipper, S.P., Howard, L.A., Houghton, G., 1998. Action-based mechanisms of attention, *Philosophical Trans.: Biol. Sci.*, 353(1373), 1385–1393.

[113] Sahu, U., Upadhyay, S., Singh, N., Patil, P., 2016. *LIBO: The Grasping Robot Using Object Recognition*, Intl. Conf. Electrical, Electronics, and Optimization Techniques (ICEEOT).

[114] Sathya, A., Anusha, A., Madhumita, P., Pavithra, R., August 2015. Systemized RFID detection of missing or misplaced books in library. *Intl. J. Emerging Technology in Comp. Sci. and Electronics (IJETCSE)*, 13(4), 235–238.

[115] Prats, M., Martínez, E., Sanz, P.J., Pobil, A.P., 2008. The UJI Librarian Robot. *Intel. Serv. Robotics*, 1, 321–335.

[116] Dahiya, R.S., Metta, G., Valle, M., Sandini, G., 2010. Tactile Sensing – From Humans to Humanoids, *IEEE Trans. on Robotics*, 26(1), 1–20.

[117] Klatzky, R.L., Lederman, S.J., Metzger, V.A., 1985, Identifying objects by touch: An expert system, *Percept. Psychophys.*, 37(4), 299–302.

[118] Allen, P.K., 1988. Integrating Vision and Touch for Object Recognition Tasks. *The International Journal of Robotics Research*, 7, 15–33.

[119] Okamura, A.M., Cutkosky, M.R., 1999. *Haptik exploration of fine surface features*, IEEE Int. Conf. on Robotics and Automation (ICRA).

[120] Omata, S., Murayama, Y., Constantinou, C.E., 2004. Real time robotic tactile sensor system for the determination of the physical properties of biomaterials. *Sensors and Actuators*, 112(2-3), 278–285.

[121] Doulgeri, Z., Arimoto, S., 2007. Force position control for a robot finger with a soft tip and kinematic uncertainties. *Robotics and Autonomous Systems*, 55(4), 328–336.

[122] Petrovskaya, A., Khatib, O., Thrun, S., Ng, A.Y., 2006. *Bayesian estimation for autonomous object manipulation based on tactile sensors*, IEEE Int. Conf. on Robotics and Automation (ICRA), 707–714.

[123] Russell, R.A., Wijaya, J.A., 2003. *Object location and recognition using whisker sensors*, Australian Conference on Robotics and Automation.

[124] Russell, R.A., 2000. *Object recognition by a 'smart' tactile sensor*, Australian Conference on Robotics and Automation.

[125] Schneider, A., Sturm, J., Stachniss, C., Reisert, M., Burkhardt, H., Burgard, W., 2009. *Object Identification with Tactile Sensors using Bag-of-Features*, IEEE/RSJ Intl Conf on Intelligent Robots and Systems (IROS), 243–248.

[126] Bierbaum, A., Rambow, M., Asfour, T., Dillman, R., 2008. *A Potential Field Approach to Dexterous Tactile Exploration of Unknown Objects*, IEEE-RAS Intl Conf on Humanoid Robots, 360–366.

[127] Gorges, N., Navarro, S.E., Göger, D., Wörn, H., 2010. *Haptic Object Recognition using Passive Joints and Haptic Key Features*, IEEE Intl Conf on Robotics and Automation (ICRA), 2349–2355.

[128] Liu, H., Yu, Y., Sun, F., Gu, J., 2016. Visual-Tactile Fusion for Object Recognition. *IEEE T Autom. Sci. Eng.*, PP (99), 1–13.

[129] Kus, Z., 2016. *Dynamical Pattern Vector in Pattern Recognition with the Use of Thermal Images*. 8th Intl Conf Computer and Automation Engineering (ICCAE).

BIOGRAPHICAL SKETCH

Ester Martinez-Martin

Affiliation: Universitat Jaume I (UJI)

Education: PhD

Ester Martínez-Martín is member of the Robotic Intelligence Laboratory at the Universitat Jaume I (UJI Castellón). She holds a B.Sc. in Computer Engineering (2004) and a Ph.D. in Computer Engineering (2011), both from Universitat Jaume-I. Her education is completed with postgraduate certificates in different topics such as computer design, programming languages and web design.

She has been involved in robotics research, being part of several national and international projects (e.g., FP7 Eyeshots (Heterogeneous 3D Perception across Visual Fragments), and FP7 GRASP (Emergence of cognitive grasping through emulation, introspection and surprise)). Her research background has been extended by participating in several relevant international conferences and schools, as well as with research stays at University of Genoa, SungKyunKwan University and University of Vienna. She is author and co-author of relevant scientific publications – including the book "Robust motion detection in real-life scenarios" (Springer 2012) – and editor of some books. She has been the Organization Chair of the IEEE-RAS Summer School on Experimental Methodology, Performance Evaluation and Benchmarking in Robotics (2015), 13th International Conference on the Simulation of Adaptive Behavior (SAB 2014), and 12th International UJI Robotics School on Perceptual Robotics for Humanoids (2012), and has collaborated in some conference committees and outreach activities. She recently received the Santander Award for Young Researchers in the field of engineering and architecture.

Angel P. del Pobil

Affiliation: Universitat Jaume I (UJI)

Education: PhD

Angel P. del Pobil is a professor at Jaume I University (Spain), where he was the founding director of the UJI Robotic Intelligence Laboratory. He has been Co-Chair of two Technical Committees of the IEEE Robotics and Automation Society and is a member of the Governing Board of the Intelligent Autonomous Systems (IAS) Society (2012-present) and EURON (European Robotics Research Network of Excellence, 2001-2009). He has over 250 publications, including 13 books, the last three published recently by Springer. Prof. del Pobil was co-organizer of some 50 workshops and tutorials including: five at ICRA (1996, 2000, 2010-2012), ten at IROS (2000, 2004-2013), three at RSS (2008-2010, 2012), ECAI'04, ICAR'05 and ACM/IEEE HRI 2010. He was Program Co-Chair of the 11th International Conference on Industrial and Engineering Applications of Artificial Intelligence, and General Chair of the 13th International Conference on Adaptive Behaviour (SAB 2014) and of five editions of the International Conference on Artificial

Intelligence and Soft Computing (2004-2008). He is Associate Editor for ICRA (2009-2015) and IROS (2007-2012) and has served on the program committees of over 130 international conferences, such as IJCAI, ICPR, ICRA, IROS, EUROS, IAS, ICAR, CIRA, etc. The UJI Robotic Intelligence Lab has organized 12 consecutive editions of IURS, the International UJI Robotics School. He has been General Chair for IURS 2005 on Robotics and Neuroscience, IURS 2006 on Humanoid Robots, IURS 2009 on Visuomotor Coordination, IURS 2012 on Perceptual Robotics and the 2015 IEEE-RAS Summer School on Experimental Methodology, Performance Evaluation and Benchmarking in Robotics. He has been involved in robotics research for the last 30 years. Professor del Pobil has been invited speaker of 63 tutorials, plenary talks, and seminars in 14 countries. He serves as associate or guest editor for 12 journals, and as expert for research evaluation at the European Commission, the US National Science Foundation and the Spain Ministry of Science. He has supervised 16 Ph.D. Thesis, including winner and finalists of the Georges Giralt EURON PhD Award and the Robotdalen Scientific Award Honorary Mention. He has been Principal Investigator of 30 research projects. Del Pobil is an active member of the IEEE Robotics and Automation Society and a lifetime member of the Association of the Advancement of Artificial Intelligence (AAAI).

In: Robotics: New Research
Editor: Daniel G. Fisher

ISBN: 978-1-63485-967-7

Chapter 2

DETECTING AND TRACKING HUMAN MOTION WITH VISION BASED ROBOTICS

Mahmoud Tarokh*
Department of Computer Science, San Diego State University,
San Diego, CA, US

ABSTRACT

The chapter discusses recent progress in vision-based robotic person following. After briefly explaining the developments in the past decade, several approaches to person detection and following are described. Color based person detection in image using methods of color tolerance, color ellipsoid and color histogram are described. The effects of light changes on color-based person detection are explained and a method to compensate for significant environment light changes on the person's image is proposed. The camera internal gain and shutter speed controls enable person following when the person move from bright (sunny) areas to darker locations (shade) and vice versa. Region shape characteristics provide complementary information to color criteria for person identification, and a number of shape measures are suggested. In addition, a technique based on edge detection and Hough transform is described that allows color-independent person detection to complements the color dependent detection. In order to reduce computational burden and achieve

* Corresponding author email: mtarokh@mail.sdsu.edu

real-time person following, a method is described to limit the area in the image that is searched to locate the person. This area is expanded or contracted adaptively based on detection confidence of the previous frames. In order to follow the person while keeping a reasonably constant distance from it, the robot speed and steering, and the camera pan and tilt are controlled. The control loops use information about the area and centroid of the detected upper body, e.g., torso or head in the image. Experiments are conducted using two robotic platforms; a Segway RMP and an ActiveMedia All-terrain rover. Experimental results are reported in various indoors and outdoors environments.

Keywords: person identification, robotic person following, robot control

1. Introduction

Over the past 15 years there has been considerable research in the area of robotic person following. Tracking and following a person by a robot has many applications such as surveillance, search, rescue, combat and security. Such a robot can also be used as a human assistant for carrying tools and equipment, and transmitting person's images to a central station. A major requirement in these applications is the ability to track and follow a moving person through non-predetermined, unstructured and sometimes rough environments. The main task in vision based robotic person following is identifying the person, and providing information on his location, size, etc. A secondary task is to control the robot based on this information to follow the person. The majority of detection methods have concentrated on vision using color images. However a number of other sensors, especially laser ranger finders have also been employed.

There are several approaches to robotic person following, namely, color matching, optical flow, stereo-vision matching, feature extraction, just to mention a few. A simple segmentation method combined with certain shape measures has been developed in [1]. This system, however, has limited capabilities due to the use of gray scale images and fixed camera with no pan and tilt. A color based tracking system capable of tracking color blobs has been implemented in [2]. Color histogram is employed in [3] to locate the person using two independently moving cameras with triangulation to determine the person to robot distance. However, the system in [3] relies on the lighting to be constant and is prone to difficulties in non-straight (turning) person movements. Another vision based system is described in [4] which uses certain feature parameters such as position, color and speed of the motion of detected region in the image. However, specifics of the detection technique are not disclosed. In [5] cues about human behavior during walking are utilized to improve reliability of person following. However, the reported results are for a simplified indoors environment. More recently, color and feature identification with fuzzy logic control was proposed for following a person in unstructured environments with reasonably constant lighting [6]. The method's performance degrades in outside environments when there are significant light changes.

Optical flow methods have been reported in [7]-[10]. In [8] velocity information is extracted from the image optical flow, whereas in [9] the person is segmented from the background assuming motions of the person and background are different. Optical flow techniques suffer from drift, especially when the person moves on non-straight paths.

Stereo-vision matching is discussed in [11], [12]. A dense stereo matching is described in [11] that reconstructs a two-dimensional view of the objects in the environments. It uses wheel odometry to estimate and subtract the motion of the background. The location of person is obtained from the resulting motion using a Kalman filter. A major problem with this approach is its dependence on odometry which is unreliable when the robot moves over uneven terrain as well as various wheel slips that adversely affect the results. Indoor person detection based on sparse features using grey-scale stereo cameras is reported in [12]. The features are matched between stereo images and are tracked over time. Methods for tracking features have also been proposed in [13]-[15].

Zajadel et al. propose a Bayseian algorithm which uses color cues for tracking humans when they disappear then reappear in the scene [16]. The method is sensitive to illumination changes and is applicable for indoor tracking with very few persons in the scene. Since detection based on color is sensitive to light, Tarokh and Merloti [17] propose to control camera internal and external parameters automatically to keep the camera light exposure constant, and significantly reduce the effect of illumination changes during difficult walking maneuvers. To deal with the opposite phenomenon, i.e., low lighting, Sales et al. [18] use multiple sensors consisting of vision system, sonars and laser rangefinders. Chen and Birchfiled [19] use binocular vision and Lucas-Kanade feature detection to determine the location of the person in the image where matching is performed between two images of a stereo pair, as well as between successive video frames. However the method relies on sparse features which make it distracted by other moving object. Person identification using not only clothes but also face images [20] or gait patterns [21] have proposed. However, person following cannot work with face or gait information.

In order to reduce the dependency on color of person identification, Tarokh and Shenoy [22] employed Canny edge detection and Hough transform to locate the person's head and then searched for the torso under the head. This combination of color independent head detection followed by color dependent torso detection was found to be effective in person identification. Other researchers have proposed illumination invariant descriptors to make the task of recognition independent of light variations (e.g., [23] and references therein). However, these approaches rely on many restricting assumptions such as homogeneity, co-planarity, multiple training under varying illuminations, none of which is applicable to the person following situations. In fact, it has been proven in [23] that there are no discriminative features of objects that are illumination invariant.

Reference [24] discusses a promising combination of an omnidirectional camera and a laser with the latter as the primary sensor. Blobs resulting from the laser scans are matched against regions in the image and a probabilistic approach is used to determine the position of the person. The work in [25] uses global positioning system (GPS) and wireless to transmit the position of the person to the robot. This method, however, relies on the accuracy of the GPS which is generally poor when higher accuracies are needed, i.e., the when robot is required to closely follow the person.

Histogram of oriented gradient (HOG) [26] has shown success in detecting people. The detection method has also been extended to partial occlusion [27]-[28]. However, the processing speed and real-time application, which are essential for person following, have not been the main consideration. In addition a combination of HOG and distance information using RGB camera and depth measurement (RGB-D) with Microsoft Kinect has been proposed [29]-[30]. Experiments in indoor environments with people around using RGB-D have also been reported [31]. Kinect sensor, however, cannot be used in outdoor environments, especially under sun light. Ess et al.[32]-[33] proposed to integrate various cues such as appearance-based object detection, depth estimation, visual odometry, and report good results. However, the implementation is costly to be used for controlling a real robot.

Laser Range Finders (LRF) are an important part of people detection and tracking systems in surveillance and robot and human interaction systems. Laser range-finders in conjunction with vision have been used in robotics person following schemes. LFRs are important in people detection and tracking systems in surveillance and robot and human interaction systems. Due to safety regulations, applications using lasers are mostly limited to lower human body. Thus legs have been widely used for human detection and tracking, e.g., [34]-[35]. When compared to other sensors (e.g., vision), a limitation of LRFs is that range data is not enough to distinguish physically different objects when they are in contact (e.g., a person leaning against a wall). To deal with this problem Carballo et al. [36] propose a multi-layered people detection using four LRF sensors that are installed in a mobile robot, arranged in pairs and compute a set of features to detect and fuse body parts.

More recently scale-invariant feature transforms (SIFT) proposed by Lowe [37] has been used for person identification in person following. [38]. In this work, the robot is equipped with cameras and LRF, and a method is proposed for identifying a person based on the pattern of clothing using the SIFT. A set of distance-dependent threshold are used to cope with the decrease of the number of feature matches due to the increased distance. Another feature detection method, speeded up robust features (SURF) [39] has found application in wheelchair following [40]. This work is based on Microsoft Kinect RGB-D cameras and SURF algorithm is employed to provide description features of the environment and target. The limitations of these methods for person following are decrease in the number of features with increased distance, and the cost of LRF.

In this chapter we discuss vision based person following, and explain several techniques for person identification in an image for real time robotic person following. Results of experiments are also provided.

2. Moving Person Detection Using Color

The first task in a person following scheme is the detection and segmentation of the person from the scene. This task consists of two subtasks, namely, training a detection system and recognition of the person by such characteristics as color, size and position in the field of view as he/she moves in the environment. Color plays an important role in the recognition task. Below we discuss several color based person recognition methods.

For training of the system, the person appears in front of the camera at the start of a tour, and images of the person are captured automatically when the person takes several poses, i.e., back to camera, and side views. The system is then trained to recognize the color of the person's upper body (torso/shirt). There are a number of color descriptions, the H (hue or color), S (saturation or color depth), I (intensity) color model or HSI is based on direct interpretation of colors and provides a better characterization compared to other color models such as RGB for this application. There are several methods for color matching, two of which are described below. The first is color tolerance matching, and the second is color histogram matching.

2.1. Color Tolerance Matching

This method applies to cases where the color of the person's shirt is reasonably uniform. In this method, the averages of H, S and I components of the torso (shirt) for the poses mentioned above are recorded, which provide the nominal values H_{nom}, S_{nom} and I_{nom}. However since these values will go through changes during the motion, we allow tolerances ΔH , ΔS , ΔI from the nominal values, which are found experimentally. These tolerances are taken as standard deviations values. Thus during the person following, if a pixel in the image has color components within the reference values $(H_{ref} = H_{nom} \pm \Delta H)$, $(S_{ref} = S_{nom} \pm \Delta S)$ and $I_{ref} = I_{nom} \pm \Delta I$, then the pixel will be a candidate for belonging to the person's image.

During person following, the camera takes images of the scene and the system performs several operations to segment the person from other objects. The first operation is to scan every pixel and mark the pixel as belonging to the person image, e.g., set it is to white if all its three color components are within the reference color ranges H_{ref}, S_{ref} and I_{nom}. The process of checking all pixels is time consuming, and therefore it is speeded up by considering two observations. First, since the person's image occupies a large portion of the image, it will be sufficient to check pixels on every other row and every other column for color verification. This way only a quarter of the pixels are checked and marked if they satisfy the color tolerance. The skipped pixels will be marked if the checked pixels around them have been already marked. The second observation is that there is a maximum distance that the person can move between two consecutive frames. As a result, the person's pixels in the current frame must all lie within a circle centered at the centroid (to be defined shortly) of the previous frame. These two observations limit the number of pixels to be checked and speed up the marking of the pixels that belong to the person's image. The final operation is to perform a standard region growing to the collect marked pixels to form connected regions.

Figure 1 (upper left) shows an example of the scene, and Figure 1 (upper right) shows the identified regions after pixel color components are checked and marked white, and region growing is applied. Note that color checking is performed only on the pixels inside the circle shown and described above. In this example three regions have been identified that have satisfied the color matching. Regions smaller in area than a specified value are considered noise and are removed as evident in Figure 1 (lower left). In this example only two candidates regions remain as a result of color checking, i.e., regions 0 and 1 in Figure 1(lower left). The smaller region, marked as 1, will be removed after the shape matching, to be described later.

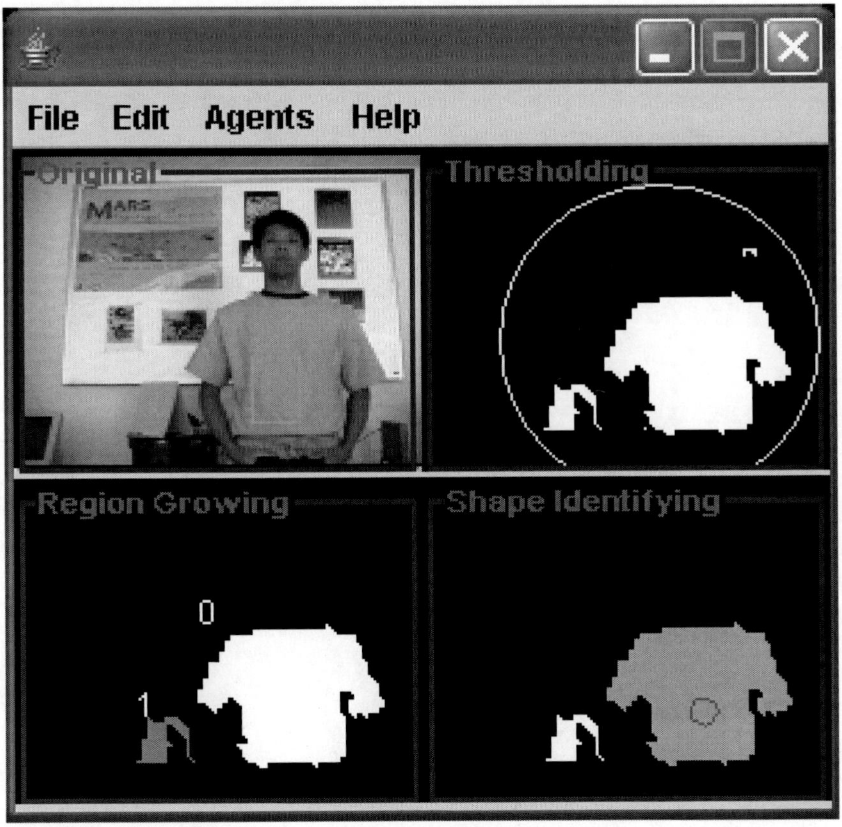

Figure 1. Person detection with color thresholding and region growing.

2.2. Color Ellipsoid Method

An alternative and improved method of color based thresholding and accepting pixels belonging to the person's torso is as follows. The colors in the selected area are converted to the HSI model. Note that pixels selected from an even uniform color shirt will have close but different HSI values. We consider only hue and saturation to reduce the dependence on lighting. For a shirt with reasonably uniform color, there may be a couple of hundred different hue-saturation pairs (h_i, s_i), $i = 0,1,\ldots m$ where m is the number of pairs in the selected rectangular area. Color variation in the rectangular area can be represented by standard deviations of the set (h_i) and (s_i), denoted by Δ_h^2 and Δ_s^2, respectively. Each pair (h_i, s_i) forms a point in the H-S plane, with a tolerance represented as shown in Figure 2. Thus each point in the H-S plane is replaced by a small ellipse given by

$$\frac{(h-h_i)^2}{\Delta_h^2} + \frac{(s-s_i)^2}{\Delta_s^2} = 1 \; ; \; i = 0,1,\ldots m \tag{1}$$

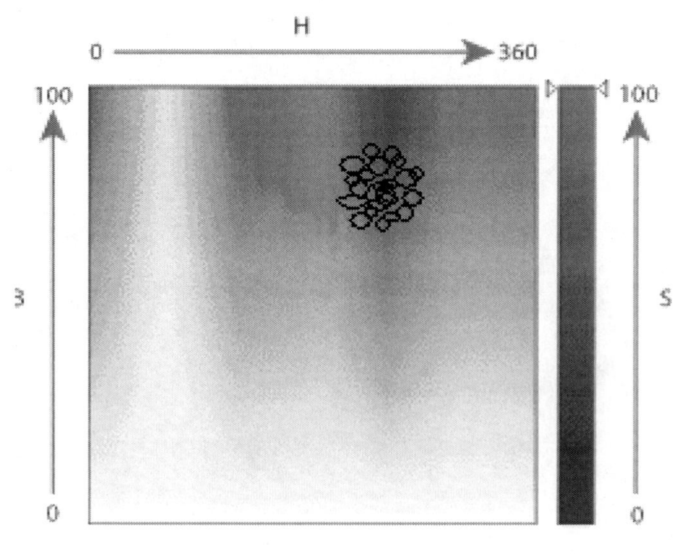

Figure 2. The hue-saturation plane showing the deviation of points with ellipses.

During person following, pixels in the image that fall inside one of the ellipses described by (1) will be considered admissible and will be candidates for further processing. The advantage of the above method over the color tolerance method is that instead of individual components, combined hue and saturation is used to decide on the pixel for belonging to the person's torso.

2.3. Color Histogram Matching

The above two methods find candidate pixels belonging to the person's image in simple cases when the area of interest (torso/shirt) has a reasonably uniform color. They fail when color variations are moderate or large. A better method of matching is using the histograms of color components. In this method the color histogram of the torso is found during the training phase. During the person following phase, the color histogram of the candidate region is found. This histogram is compared with the histogram of the training.

We use only histograms of hue and saturation for comparison to reduce the effect of light changes on the results, as explained before. Usually hue ranges between 0 and 360 degrees on the color wheel, and saturation ranges between 0 and 100%. Let there be L_1 values of hues, and L_2 be values of saturations. For the purpose of comparison of two normalized histograms during training and following, it is convenient to combine hue h and saturation s into a one-dimensional normalized histogram array $A(q),\ q = 1,2,3,\ldots,L$ where $L = L_1 + L_2$, the first L_1 values are for the hue, the next L_2 are for saturation and $0 \le A(q) \le 1$ due to normalization. Let the normalized histogram of training and following be denoted by $A_t(q)$ and $A_f(q)$, respectively. In order to compare the closeness of the person following histogram to the trained histogram, we use Bhattacharya coefficient which is defined for the distance between two probability distributions. Since a histogram is a essentially a probability distribution, we use the Bhattacharya coefficient for the similarity measure between the two normalized histograms as

$$\sigma = \sum_{q=1}^{L} \sqrt{A_t(q)\, A_f(q)} \tag{2}$$

When the two color histograms are identical, $A_t(q) = A_f(q)$; $q = 1,2,3,\ldots,L$; $\sigma = \sum_{q=1}^{L} A_t(q) = 1$. Furthermore, $\sigma = 0$ when they are completely difference that is when $\{ A_t(q) = 0, A_f(q) \neq 0 \}$, or $\{ A_t(q) \neq 0, A_f(q) = 0 \}$ for all q values.

2.4. Effects of Lighting on Color

The methods described above reply on the assumption that colors do not change appreciably with changing light during the person walk. The color components, however, do vary with light intensity, and even in the HSI color model, hue and saturation are not entirely decoupled from intensity. When a person moves from a shade to a sunny area, the light intensity varies substantially (as much as two orders of magnitude) and hue and saturation also vary considerably.

Experiments have shown that for each color there is a range of light intensities where hue and saturation are reasonably constant. As a result, for each color we define a desired light intensity I_d that corresponds to the middle value of the range for which hue and saturations remain essentially constant for that color. The desired intensity I_d is used as the reference or the set point during the training for the camera light exposure control.

The image intensity I_i produced by a camera is a function of three parameters, namely, camera gain *cg*, shutter speed *ss* and iris radius *ir*, that is $I_i = f(cg, ss, ir)$. This function is nonlinear, but is linear for a limited range. The gain changes the sensitivity in a CCD device; higher values of gain produce higher image light intensities. The shutter speed allows adjusting the duration of the light that passes through the camera. Finally, iris changes the area of the diaphragm through which the light passes to the camera. Lower shutter speeds causes image blur, and therefore the goal would be to adjust the gain and iris for the desired intensity while keeping the shutter speed closer to its upper range.

Let the desired light intensity mentioned above for the person's torso/shirt be I_d, then the intensity error is $I_e = I_d - I_i$ where I_i is the image intensity of the person's torso. The image intensity is determined by a mask containing a "software sensor". The sensor consists of a larger square in the middle and four smaller squares around it, as shown in Figure 3. The sensor is placed on the target image during person following. The pixels under the larger center square and four smaller squares are read and their intensities are averaged to obtain the measured I_i . Using this procedure the intensity of the representative pixels of the target is determined.

(a) (b)

Figure 3. (a) The image produced by the camera self-exposure adjustment in a dark environment, (b) the image as a result of control of camera parameters in the same environment.

A two level controller acts on the error I_e to achieve the desired intensity I_d . Because of significant difference between indoors and outdoors lighting, the camera iris is initially set to a high opening for indoors and a low opening for outdoors operation. The shutter speed is initially set to a high value since lower shutter speeds can cause blurring, as mentioned before. The first level control is the gain adjustment so as to reduce the light intensity error to a minimum. The shutter speed is reduced by the controller only if automatic adjustment to gain does not achieve a low error I_e . The controller for the light intensity is a standard proportional plus derivative type and the controller parameters are tuned empirically to obtain a fast response. Typically the desired light intensity is achieved after 3 to 10 frames in our experiments which had a processing speed of 15 to 20 frames per second.

It is noted that cameras also have self-adjusting exposure. However, the adjustment is often based on the average intensity of the whole image rather than the area of interest (person's torso in our case). Figure 3(a) shows an image in a dark room using the camera self-adjusting exposure, which has resulted in dark color for the target. Figure 2 (b) shows the same environment where the camera parameters are adjusted by the proposed two-level controller using the desired intensity. It is seen that the hierarchical controller has brought the target color to its correct value. Some cameras allow self-windowed exposure control, which is an improvement over entire image exposure adjustment. However, we still need our own exposure control in order to be able to adjust the light intensity based on the trained value. Without light intensity control, the person following fails in environments with intensities different from those of the training environment.

3. Region Shape Detection

Detection of pixels through color thresholding produces a set of pixels that are candidate for belonging to the region of interest (target), i.e., the person upper body (torso). However these pixels do not usually form a connected region, and may form several collections of pixels some of which may not be on the target due to errors in thresholding. As a result, standard region growing in applied which generally results in a number of regions. In what follows we discuss several methods for identifying the correct target.

3.1. Torso Detection by Shape Measures

A shape identification system can be trained using two poses of the person, one with the back of the person to the camera and other the person's side view, i.e., when the person turns about 90 degrees. Shape measures must be independent of the mass (area) of the person's image because the mass changes with robot to person distance. The three measures that satisfy this requirement are compactness, circularity and eccentricity. The circularity C measures the closeness of the region shape to a perfect circle having a radius equal to the average of distances of center of a region to the boundary points (pixels). The compactness Q is the ratio of area to perimeter of the region (torso) raised to the power of 2. Finally, eccentricity E is the ratio of the minimum and maximum distance of the center of the region to its the boundary. The normalized values of the three measures are between 0 and 1. During the training, each of these measures is evaluated for the person in each of the above two poses (k = 1,2) and their values $C_{k,ref}$, $Q_{k,ref}$ and $E_{k,ref}$ are stored for the person following phase. This completes the training of the system, which can be considered as an off-line phase.

After region detection, any region smaller in area than a specified value, is considered noise and is removed, as mentioned before. The shape measures C_i, Q_i and E_i for the remaining candidate regions are computed, where $i = 0,1,2,...,m-1$ denotes the region numbers. Rather than checking each shape parameter with its corresponding reference value, we define a single measure for the closeness of the detected region to the reference region which is the person's image found during the training. A possible measure is

$$\beta_{k,i} = \frac{1}{3}\left(e^{-a_C \delta C_{k,i}} + e^{-a_Q \delta Q_{k,i}} + e^{-a_E \delta E_{k,i}} \right);\ k=1,2;\ i = 0,1,2,...,m-1 \qquad (3)$$

where $\delta C_{k,i} = \frac{|C_{k,ref} - C_i|}{C_{k,ref}}$, $\delta Q_{k,i} = \frac{|Q_{k,ref} - Q_i|}{Q_{k,ref}}$, and $\delta E_{k,i} = \frac{|E_{k,ref} - E_i|}{E_{k,ref}}$ are the relative deviations from the reference shape parameters, and a_C, a_Q and a_E are user specified coefficients that determine the relative weighting of each shape measure to the closeness measure. The exponential function provides high sensitivity to mismatch between the region and the reference shape. Note that the particular closeness function chosen in (3) produces 1 if all shape measures of the region are the same as the reference value, and approaches zero if the region shape measures are completely different. It is noted that for each detected region, two shape measures are found, i.e., one at each pose. The region that has the largest value of β is selected, and if this value is close to 1, the selected region is assumed to represent the person. If all the regions have small values of β, then none is chosen and another image is taken and analyzed. In Figure 1 (lower right), the larger region is the one that matches the shape measure of the person.

3.2. Color Independent Head and Torso Detection

So far the person detection has been described based on color thresholding, and region growing. In this section we describe edge detection, edge linking by the Canny's method and a modified Hough transform to find the region of interest for the person detection. Human head has a circular shape and therefore the head can be used, in addition to torso, for distinguishing the person from other objects in the image.

In this subsection, we briefly explain the well-known Canny edge detector as is applied to our problem. First Sobel edge operators in x and y directions are applied to the top third of the image, and the gradient magnitude is found. Note that only a rectangle in the upper part of the image is considered for detection of the head. Limiting the search area for the head detection speeds up the operation and reduces false detection, as will be discussed. Later we will discuss a more intelligent method to adaptively expand or shrink the region where the head is to be searched.

Non-maximum suppression, or thinning of edges, follows Sobel edge detection. The suppression is basically a search to determine if the gradient magnitude assumes a local maximum in the gradient direction and if such a point exists, that single point is selected across the width (thickness) of an edge.

The next operation is thresholding to get strong edges and to discard weak ones. For better results hysteresis thresholding is applied which exploits the fact that edge pixels are normally connected. This solves the challenge of choosing a single appropriate threshold value. For hysteresis thresholding, two thresholds are chosen, a high T_h and a low T_l. Any edge pixel with strength (gradient magnitude) greater than T_h is retained, and any edge pixel with strength less than T_l is discarded. An edge pixel e with strength between T_l and T_h is retained only if there is a path of edge pixels with strength greater than T_l connecting e to another edge pixel e' of strength greater than T_h. When this step is completed we are left with a binary image with each pixel being marked as either an edge pixel or a non-edge pixel.

3.2.1. Head Detection

After edges are detected as described above, Hough transform is applied to find the circular shaped head. Hough transform is a method for finding position of simple shapes hidden in large amounts of other data. Since the equation of the circle is known, no prior knowledge of its position is needed. The main advantage of Hough transform is that even if part of the object's edge is available (visible), the object can be detected in most cases. Below we describe a modified Hough transform method for the circle (head) detection.

A circle is specified by three parameters, namely its radius R and its center coordinates $(\hat{x}, \hat{y})$. It is noted that the nominal person's head radius R in the image is determined at the training phase, where the person stands at a fixed distance from the camera mounted on the robot and his/her head size (radius) in the image is recorded. During the person following phase, the robot speed control system maintains this desired distance, as will be explained later. Thus the head size (radius) in the image is relatively constant. As a result we only need to find the center $(\hat{x}, \hat{y})$ to locate the head in the image.

It is noted that the center of the circle $(\hat{x}, \hat{y})$ can be at any location within the search area (window) for the head. Let the height and width of the search window be denoted by *H* and *W*, respectively. Initially, the search window is relatively large, e.g., the top third of the image, but is then adjusted, as will be explained in the next section. In order to determine the most likely location for the circle center, we divided the search window into an *m* by *n* grid by dividing *W* into *m* intervals and *H* into *n* intervals such that $\alpha = \frac{W}{m} = \frac{H}{n}$. If needed, the width or height of the search window is increased slightly so that both *m* and *n* are integers. This produces $m \times n$ squares each with the side length of α. The value of α determines the granularity of the division, and is fixed. Let the coordinates of the center of the squares be denoted by (x_i, y_j), $i = 1,2,\cdots,m$; $j = 1,2,\cdots,n$. Consider a circle centered at (x_i, y_j) and radius $r = \frac{\alpha}{\sqrt{2}}$ which passes through the four vertices of the square. These $m \times n$ circles cover the search window with some overlapping areas. The amount of the overlap is the difference in area between a circle and the square inside the circle, i.e., $(\pi r^2 - \alpha^2) = (\pi/2 - 1)\alpha^2$. We will refer to these small circles as *circular accumulators* to distinguish them from the circle representing the person's head. Using circular accumulators, instead of square or rectangular accumulators, will speed up the computation for the detection of the head, as will be seen shortly.

Consider an edge pixel located at (x_e, y_e) with a distance of d_{ij} from a circular accumulator center (x_i, y_j) which is also the center of the corresponding square. If the edge pixel is on the circumference of the circle (head) of radius *R*, the center of the circle (x_c, y_c) must be in one or more circular accumulator that satisfy

$$d_{ij} - r \le R \le d_{ij} + r, \begin{cases} i = 1,2,\cdots,m \\ j = 1,2,\cdots,n \end{cases} \tag{4}$$

Note that (4) implies that a tolerance equal to r on the trained head radius R is acceptable. This is desirable since the speed control system may not able to keep the exact distance between the camera and the person, and the actual head radius in the image may be slightly different from its nominal value R determined at the training. When condition (4) is satisfied for a circular accumulator, its count C_{ij} is set to one. This procedure is performed for all edge pixels, and if (4) is satisfied, the corresponding count C_{ij} is incremented by one. At the end of this procedure the accumulator with the highest count indicates the likely circular accumulator that contains the head center. Thus the center of the accumulator with the highest count will be considered as the head center $(\hat{x}, \hat{y})$. However, it can happen that a circular shaped object other than the head in the search window to produce a higher count. To account for such cases, we consider several (typically three) highest count accumulators, instead of only the highest. Torso that is below the head will be used in the next section to select the most likely candidate for the head among the several detected circles.

It is noted that in the standard Hough transform the accumulator cells are rectangular. However, checking conditions for the center $(\hat{x}, \hat{y})$ to be inside such a rectangle is computationally more intensive than (4) and is therefore unsuitable for real-time application.

The head radius R in the image is the value obtained at the training. However, even though the robot speed control attempts to keep a constant distance between the camera and the person, this distance can change somewhat during person following. Therefore the radius of the detected head is adjusted based on the average of the last several values. This updating has been found to be effective in estimating the correct circle radii for the detection of the head.

3.2.2. Torso Detection

In order to substantially reduce the chance of false head detection, we search for the person's torso under each detected circle (potential head) in the image.

The torso shape is somewhere between a rectangle and an ellipse. A mathematically convenient shape for the torso is the super-ellipse which is described by

$$\left(\frac{x}{a}\right)^p + \left(\frac{y}{b}\right)^p = 1 \tag{5}$$

where a and b are the width and height of the ellipse, respectively and $p > 2$. Higher values of p produce shapes similar to a rectangle with rounded corners while values of p close to 2 produce a shape similar to an ellipse. It is found that the value of p between 2.5 and 3 is suitable for representing the general shape of a person's torso. The nominal values of a and b are found during the initial training when the person stands with back to the camera at a fixed distance.

We adopt a procedure similar to that of circle detection. Let us assume again that the search window under the detected head is divided into a grid, and consider circular accumulator cells with radius r. Consider an edge pixel (x_e, y_e) on the circumference of the supper-ellipse, and let the center of an accumulators cell be denoted by (x_i, y_j), as before. Unlike the circle where the distance between its center and a point on its circumference is constant equal to R, the corresponding distance for a super-ellipse can be shown to be

$$S = \sqrt{a^p \cos^2 \varphi + b^p \sin^2 \varphi} \tag{6}$$

where $\sin\varphi = \left(\sqrt{\frac{a}{b}}\right)^p$ for $b \geq a$ which is the case for the torso. Note that when $a = b$ and $p = 2$, we have a circle and $S = R$ in (4).

The accumulator cells are constructed as in the case of head detection and the distance d_{ij} between an edge pixel (x_e, y_e) and the accumulator center (x_i, y_j) is found as before. The accumulator count denoted by E_{ij} is incremented if the following condition, which is identical to (4) with radius R replaced by S, is satisfied

$$d_{ij} - r \leq S \leq d_{ij} + r, \quad \begin{cases} i = 1,2,\cdots,m \\ j = 1,2,\cdots,n \end{cases} \tag{7}$$

Finally the center of the accumulator with the highest count is chosen as the center of the supper ellipse.

As a result of the above operations a super-ellipse is found under each potential (candidate) head location. Let the highest accumulator count for three circles representing a potential head be C_1, C_2, C_3, respectively. Let the corresponding maximum accumulator counts in the super-ellipses under the three circles be E_1, E_2, E_3, respectively. The normalized counts for the circle and ellipsoid are now defined as

$$c_\ell = \frac{C_\ell}{\max_\ell (C_\ell)}, \quad 0 \le c_\ell \le 1; \quad e_\ell = \frac{E_\ell}{\max_\ell (E_\ell)}, \quad 0 \le e_\ell \le 1; \quad \ell = 1,2,3 \tag{8}$$

Then the head circle/torso super-ellipse that provides maximum weighted sum described by the following equation will be the one chosen to represent the person

$$\mu = \min_i \{ w c_\ell + (1-w) e_\ell \}; \; \ell = 1,2,3 \tag{9}$$

where $0 \le w_\ell \le 1$ is a weighting factor that emphasis or deemphasizes the head count relative to the torso count. Equation (9) will be referred to as the *matching criterion.*

Although the above scheme is effective for following the person during walks on straight paths, it faces difficulties when the person makes sharp turn in which case the torso is seen as a narrow super-ellipse. In this case, the width of the super-ellipse a becomes small and differs from the trained value. A solution is to use the value of the camera pan to determine the amount of the turn and adjust a accordingly. This improves the detection of the torso during person's turns, but adds slightly to the computation time.

4. Limiting and Adapting Search Area

It is essential that during person following, many frames per unit time be processed to keep track of the person in real-time. This requires considerable computation if all image pixels are processed. As a result, rather than processing the whole image for region detection, the search area in the image must be limited for locating the target, which is either the torso in the color based detection or head in the edge and Hough detection. Search area is defined as a rectangular region that is required to be processed in order to locate the target (torso or head). The smaller the search area, the faster is the computation and therefore the faster the person can move and be followed in real-time. Not only real-time following becomes possible when limiting the search area, but also the chance of false target detection becomes lower since there are fewer objects similar to the target in a smaller search area. Therefore adaptation of the search area is very important.

4.1. Search Area Adaptation

For color based detection, the candidate area for locating the person in the next frame is restricted to be: (i) Near the center of the image (origin of the x-y plane), (ii) Near the last known location of the person, (iii) Everything between the above two and (iv) Biased by the direction of motion

The restriction (i) is due to the fact that the camera pan and tilt controls bring the target to the center of the image, as will be seen. Restriction (ii) accounts for the fact that due to the short time interval between the capture of two consecutive frames, the target does not move appreciably from its previous location. The limitation (iii) above is indicative of uncertainty when items (i) and (ii) are significantly different. The centroid (x_c, y_c) of the detected region is important and is used for controlling the robot as will be seen in Section 5. Finally, the change in centroid $(\delta x_c, \delta y_c)$ of the last few frames provides the direction of the motion and is useful in estimating the possible location of the person in the next frame. The white rectangular boundaries in Figure 4 show example of expanding or contracting the areas used for detection of the target. Figure 4(a) shows contracted search area when the target (person in blue shirt) was detected in the previous frame and Figure 4(b) shows expanding the search area when the target is lost.

In order to determine the size and location of the search rectangle, we use the centroid of the current image (x_c, y_c) and the estimated change in centroid $(\delta x_c, \delta y_c)$ in the next image. Then the search area will be the rectangle bounded by

$$x_1 = x_c - \delta x_c - \frac{a}{2}; \quad x_2 = x_c + \delta x_c + \frac{a}{2} \tag{10}$$

$$y_1 = y_c - \delta y_c - \frac{a}{2} \ ; \ y_2 = y_c + \delta y_c + \frac{a}{2} \qquad (11)$$

where the a is a positive constant, and the width and the height of the search rectangle are $W = x_2 - x_1 = 2\delta x_c + a$ and $H = y_2 - y_1 = 2\delta y_c + a$, respectively.

(a)

(b)

Figure 4. (a) contracted search area when the target person (shown in blue shirt) in previous frame is detected and (b) expanding the search area when the target is lost.

During person following, we test pixel within the above rectangle to identify those whose hue h and saturation s are within one of the ellipses described by (1). This is followed by region growing to collect and connect pixels The grown region may contain "holes" due to target (person's torso/shirt) having folds and shadows that show as darker areas or patches with different colors that lies outside the trained ellipses. To overcome this problem, the boundary of the region is found using a standard method. Figure 5 shows an example of the detected regions with boundary shown in blue for illustration. The centroid of the torso (x_c, y_c), is shown in Figure 5 by a white $\times$.

It is possible that the admissibility test using (1) and the subsequent boundary detection produces several regions, as shown in Figure 5. In this figure three admissible regions have been detected, one for each upper arm and one below them. These regions have been formed due to non-uniform shirt color with a pattern in the middle.

The detected regions have different areas, and different color matching values with respect to the trained set. Suppose that r regions are detected with masses/areas M_i and centroids (x_i, y_i), $i = 1, 2, \cdots, r$. We define the normalized mass as $\mu_i = \frac{M_i}{M_{max}}$ where $M_{max} = max(M_1, \cdots, M_r)$, and $0 \le \mu_i \le 1$. It is reasonable to assume that a region contributes to the centroid of the overall detected regions proportional on its area. Thus we find the centroid of the detected regions as

Figure 5. Detection of several regions in a non-uniform shirt color and determining the centroid marked as a white ×.

$$x_c = \frac{\sum_{i=1}^{r} \mu_i x_i}{\sum_{i=1}^{r} \mu_i} \; ; \; y_c = \frac{\sum_{i=1}^{r} \mu_i y_i}{\sum_{i=1}^{r} \mu_i} \qquad (12)$$

Now consider the effect of color match on the centroid. One way to quantify the closeness of the training set colors and a detected region colors is to use color histogram described in Section 2.3. Let the color match (2) for a region be σ_i, $i = 1,2,\cdots,r$ where $0 \le \sigma_i \le 1$. We take into account the color match into the estimation of the centroid by modifying (12) as follows

$$x_c = \frac{\sum_{i=1}^{r} (\mu_i + w\sigma_i) x_i}{\sum_{i=1}^{r} (\mu_i + w\sigma_i)} \; ; \; y_c = \frac{\sum_{i=1}^{r} (\mu_i + w\sigma_i) y_i}{\sum_{i=1}^{r} (\mu_i + w\sigma_i)} \qquad (13)$$

where w is a weighting factor for the effect of regions color matching relative to region areas.

The estimated mass (area) of the image also uses the contributions of different regions based on their area and their color match with the trained region. Thus we compute the effective mass as

$$M = \frac{\sum_{i=1}^{r}(\mu_i + w\sigma_i)M_i}{\sum_{i=1}^{r}(\mu_i + w\sigma_i)} \tag{14}$$

The effective mass is a measure of the distance between the robot and the person; a higher mass corresponds to closer distance and vice versa.

Partial occlusion or person turning to the side can reduce the effective mass and also produce wrong centroid. In order to substantially reduce these errors, we use a moving average by averaging the mass values in the last several frames.

5. Robot and Camera Control

The centroid and area/mass of the detected torso or the head are used to control the robot and camera. The camera is mounted on a pan and tilt base which is attached to the robot. The objective of the robot control is to follow the person and keep a reasonably constant distance to him/her. There will be a number of control loops, two for the robot speed and steering and two for the camera for pan and tilt control. If the internal camera parameters are adjusted for light exposure, there will two additional controls for the camera gain and shutter speed as discussed in Section 2.4. The overall control scheme is shown in Figure 6.

The robot quantities to be controlled are camera pan π and tilt θ, and robot velocity v and steering ψ, as shown in Figure 7, each controlled by a PID controller. The control system obtains inputs from the vision system for the centroid (x_c, y_c) and the effective mass M of the image. The y-axis is assumed along the direction of the person motion and the x-axis is across the field of view. The origin of the coordinates system is the center of the image, which is also the desired position for the centroid of the target (person's torso). Thus the role of the pan and tilt controllers is to bring the target centroid to the center of the image. The pan controller acts on the target x_c, whereas the tilt controller acts on the target y_c, to bring both to zero. As a result the pan actuation is obtained from the standard PID controller

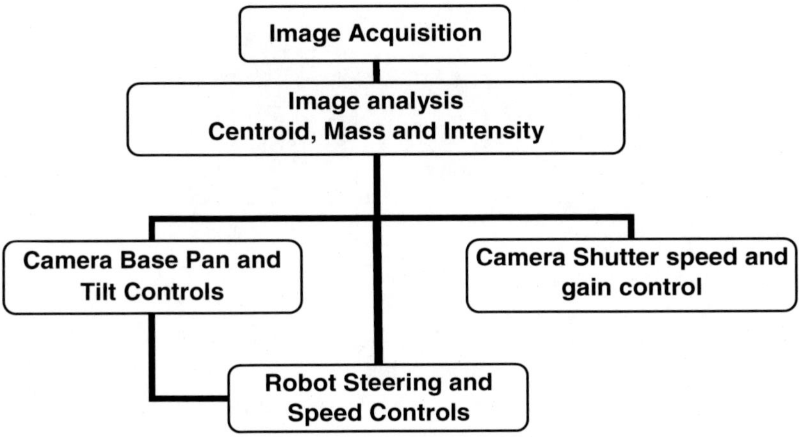

Figure 6. The overall control scheme.

The robot quantities to be controlled are camera pan π and tilt θ, and robot velocity v and steering ψ, as shown in Figure 7, each controlled by a PID controller. The control system obtains inputs from the vision system for the centroid (x_c, y_c) and the effective mass M of the image. The y-axis is assumed along the direction of the person motion and the x-axis is across the field of view. The origin of the coordinates system is the center of the image, which is also the desired position for the centroid of the target (person's torso). Thus the role of the pan and tilt controllers is to bring the target centroid to the center of the image. The pan controller acts on the target x_c, whereas the tilt controller acts on the target y_c, to bring both to zero. As a result the pan actuation is obtained from the standard PID controller

$$\pi = -g_1 x_c - g_2 \int x_c dt - g_3 \frac{dx_c}{dt} \tag{15}$$

where g_1, g_2, g_3 are controller gain parameters and the negative sign is due to the fact that the desired value of x_c is zero so that the error is $-x_c$. The controller parameters g_1, g_2, g_3 are tuned using a standard method. The tilt controller is of the form (17) except that y_c is used in place of x_c, and π is replaced by θ. The controller parameters for the tilt control are tuned similarly.

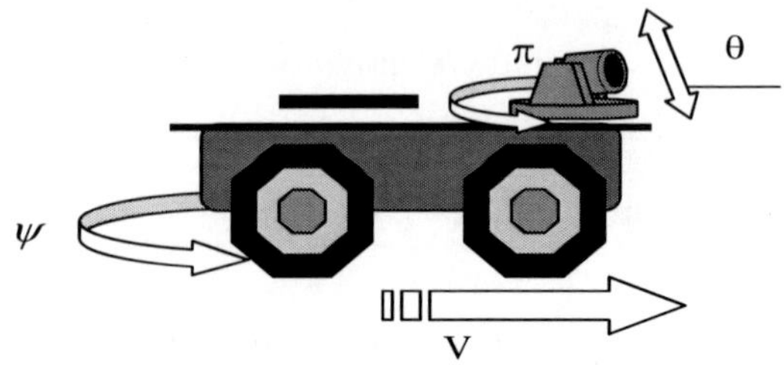

Figure 7. Robot and camera control quantities.

The camera pan and tilt behave similar to human neck, and steering behaves as the human body turn. In other words, quick and small motions of the person across the field of view are tracked by the left/right and up/down motion of the neck (camera pan and tilt) whereas larger net motion to the right or left is achieved by the body turn (robot steering). Thus the robot steering *s* is the cumulative motion of the camera pan, i.e.

$$\psi = \int \pi \, dt \tag{16}$$

The robot velocity controller uses as its reference (set point) the difference between the estimated mass obtained at the training M_{tr} and the current effective mass M of the detected person's torso. A negative mass error $(M_{tr} - M)$ corresponds to the camera (robot) being closer to the person than the corresponding distance at the training, in which case the speed is reduced. The opposite is true when the target mass is small indicating that the person is far from the robot, in which case the robot is speeded up. The robot velocity must depend on the robot turn rate so that when the robot turns, its velocity must be reduced. The following relationship captures the above reasoning and is used for the velocity set point (reference) v_{ref}

$$v_{ref} = k(M_{tr} - M)\left(1 - \frac{\varphi}{\varphi_{max}}\right) \tag{17}$$

where φ and φ_{max} are the current and maximum robot turns rates, respectively, and k is proportionality constant. Note that when $M_{tr} < M$, the velocity is negative and the robot will move backwards to keep the desired distance.

5.1. Behavior Control

In addition to following the person, the robot must avoid collision with walls and other objects, and furthermore must not come very close to the person. The latter implies that when the person stops, the robot must also come to a halt at a reasonable distance from the person. In addition, the robot must produce an appropriate action if the person is not seen by the camera, or is not detected by the image processing system. To account for these various conditions, we employ a behavior control system. There are six inputs to the behavior controller, namely, image mass *M*, image y-center of the mass y_c, average distance of robot to an object D_{ro} as measured by sonar sensors or other proximity sensors, pan π and tilt θ and a binary signal indicating if the image of the person is found. These behaviors are described as follows.

5.1.1 Search

It is possible that the image processing algorithm does not find the person for a brief period of time. This happens, for example, when the estimated mass *M* is too small or the person is out of camera view. In this case if the image is not acquired after a preset number of frames, the person is lost and the system enters a search mode behavior. This behavior starts a sinusoidal like motion of the pan and tilt of the camera with gradually increasing amplitudes so the camera moves in a spiral motion increasing the view coverage in each cycle. The equations of the pan $\pi(t)$ and tilt $\theta(t)$ are given as

$$\pi(t) = \pi_0 + ct\cos(t);\ \theta(t) = \theta_0 + ct\sin(t) \tag{18}$$

where (π_0, θ_0) are the camera pan and tilt at the time the person is lost, and *c* is a constant that determines the speed of camera motion. Typically the camera pan and tilt range is -60 to $+60$ degrees and the time allowed to find the target is 30 to 60 seconds. These values determine the constant *c*. The camera motion in the pan-tilt plane is a spiral of the form

$$\bar{\pi}^2 + \bar{\theta}^2 = c^2 \arctan\left(\frac{\bar{\theta}}{\bar{\pi}}\right) \tag{19}$$

where $\bar{\pi} = \pi - \pi_0$ and $\bar{\theta} = \theta - \theta_0$. During this behavior, the light intensity control is disabled since there is no target to control its lighting. When the target is reacquired, the intensity control becomes active again, and the search behavior transitions to the normal person following behavior. If the person is not found, a voice message to that effect is sent by the robot, asking the person to appear in the camera view.

5.1.2 Hazard

This behavior is activated if certain conditions exist on M, y_c and D_{ro}. These conditions can be, for example, that one of the three quantities (M, y_c, D_{ro}) is outside its safe range (i.e., M lager than a threshold, or y_c and D_{ro} are smaller than some limits). Alternatively, this behavior can be activated if two of the three quantities show unsafe values. The hazard behavior will result in the immediate stopping of the robot. If the problem is the robot getting close to the person as detected by the vision system (M and y_c), then new images are taken at regular intervals until the person moves again and these quantities are within their safe values. On the other hand, if the hazard is due to collision detection by the sonar sensors, then a collision avoidance maneuvering takes place. This maneuvering, can be, for example, turning away from the obstacle and then turning towards the person. Alternatively any other available collision avoidance techniques can be incorporated.

5.1.3 Normal

If the person is detected and no hazard is present, this behavior is activated. In the normal behavior, the robot will perform the person following, i.e., images are taken, mass and its center are found, and camera and robot are controlled.

6. IMPLEMENTATIONS AND RESULTS

The person following schemes described in the previous sections were implemented and tested on two robotic platforms, namely, the Active Media Pioneer 2 All-terrain robot, and Segway RMP (Figure 8). In addition tests were conducted both indoors and outdoors in different locations.

Figure 8. The Active Media Pioneer 2 All-terrain robot (left) and Segway RMP (right).

6.1. Experiments with Pioneer 2

The base dimensions of Pioneer 2 are $50 \times 49 \times 26\,\text{cm}$, has four motorized wheels, and can travel at a top speed of 1.6 m/s. The Pioneer 2 is capable of holding 40 kg and has a battery life of 10-12 hours. The robot has a sonar ring with 6 sensors, which has an operation range of 15 cm to 7 m. The sonars, shown in Figure 8 as circles, are used for obstacle detection.

The color models are 24-bit RGB and 24 bit gray scale. The laptop mounted on the base (Figure 8) is an IBM T40 with Windows XP professional operating system. It contains an Intel Centrino processor running at 1.5 MHz. The Cannon VC-C4 camera installed on the Pioneer 2 (Figure 8), permits color image capture at maximum resolution of 460 horizontal lines and 350 vertical lines. It is connected to the laptop computer through an Imperx VCE-B5A01 PCMCIA frame gabber, which is specifically designed for laptops. The frame grabber can achieve capture rates of 30 frames/second at the lowest resolution of 160×120 in NTSC format, and 5 frames per second at the highest resolution of 640×480.

The application uses a variety of software libraries written by third-party for creating interface and enabling device control. The libraries for the user interface are written in Java, whereas libraries for low motor control are in C++. As a result the person following code was written both in Java and C++. The person application uses the client server, distributed callback, model view controller. The discussion of the software is beyond the scope of the present chapter.

Extensive indoor and outdoor trials were conducted with the person following system. In these experiments significant light changes did not take place. Several scenes from indoors and outdoors experiments are shown in Figure 9. They include passage through a door, Figure 9(a), a sharp bend, Figure 9 (b), following the target as another person appears in the scene, Figure 9(c), and traversing rough terrain, Figure 9(d). These experiments demonstrated that the person following is successful in detecting the person and closely following him even when the person takes a sharp turn. Furthermore person following also takes place when the rover moves on the rough terrain.

(a) (b) (c) (d)

Figure 9. Person following in both indoors and outdoors environments.

There are two limitations with the above method, one is the number of frames per second that could be processed which was limited to 7 frames/sec, and the other is the inability of the system to cope with significant light variations.

6.2. Experiments with Segway RMP

A number of additions and modification were made to the RMP platform. The main two added components were a CCD camera and a computer. The camera is a DFK24AF from Image-Source and has a resolution of 640 x 480 and a capability of 60 frames per second. However, due to the required image processing described above, the actual processing rate is about 15 frames per second, which was found to be adequate for fast walking. Its shutter speed and gain are controllable through software while its iris is manually adjusted. The camera has a C++ class library for certain image manipulations. A pan and tilt base with servo controls and electronics was also installed for the camera. The second main component is an onboard computer. It is a mini-ITX computer with Via C7 2.0 GHz processor, 1GB RAM and 120 GB hard drive. The computer uses 12 V DC unregulated voltage which is available on the platform. It has onboard IEEE1394 port, weighs about 2 Kg and its size is 17 $\times 17 \times 7$ cm. It was installed under the top platform on a shock absorbent mount.

In addition to the camera system and computer, several other components were added to the platform. A gamepad serves for manual control of the platform, when needed, and through various buttons it allows velocity and steering adjustment. Loudspeakers were also installed on the top platform for announcing the status of the system, i.e., training, lost target, etc. Vision, tracking and control software were developed as independent modules. Standard C++ and STL were used for portability.

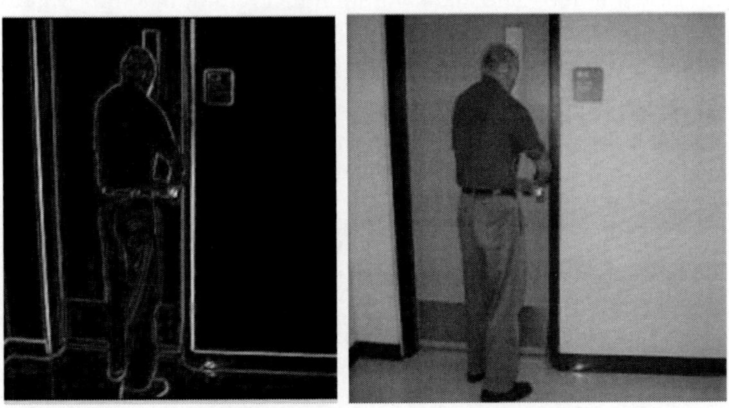

(a) (b)

Figure 10. Edge and circle detections.

6.2.1. Head Detection

The method described in Section 3.2 was applied. Figure 10 shows a typical example of the result of circle detection without further processing of the torso or restricting the search area, i.e., the search area is the whole image. Figure 10(a) is the edge detected image, and Figure 10(b) shows that two circles (in red) were detected, one for the head and another for the circle-like shape under the arm. As a result, without further processing, the correct head location cannot be determined.

Figure 11 shows the combination of head and torso detection with the supper ellipse method in three scenarios, under different lighting conditions and different person's outfit colors. It is noted that in each case more than on circular object were detected (not shown). However when a super-ellipse was searched under the each circle only the one corresponding to the torso was found/It is noted that the brightness has had no effect on the detection in this environment. It has been found that in most cases the circles-ellipse combination works well. However, when the person turns, his side instead of his back is towards the camera, the super ellipse detection method becomes problematic for the reason stated is Section 3.2.

In another experiment after the detection of three highest score circles, the region under each circle was considered for color matching, rather than a super-ellipse. Figure 12 shows typical situations during person following. At the start, the top third of the image is the search window since no prior information is available about the location of the person. A maximum of three circles were detected (only one shown in green) with three rectangular regions under them shown in red. The hue-saturation histogram matching values for the three circles were evaluated using (2). The best circle for the head is shown in green in Figure 12; the other circles with lower scores are not shown. Since the matching factor is high for the detected circle and the histogram of the region below it, the search window size (shown in blue) are expanded or contracted.

Figure 11. Head detection in different environments and with different outfits.

Figure 12. Person head detection using color matching and adaptive search.

6.2.2. Outdoor Experiments with Significant Light Changes and Walking Maneuvers

The purpose of these experiments was to evaluate the performance of the system consisting of a combination of the schemes discussed in the previous section. These include adjusting camera parameters for light exposure control, using a combination of color-based detection and color-independent detections, as well as adaptive window sizing, and pan, tilt, speed and steering controls. These experiments were performed with significant light changes and difficult walking maneuvers.

It is noted that using the Segway robot in fast walking adds challenge to person following. This is due to the forward and backward leaning motions of the robot as a result of speeding up or slowing down, though they are eventually corrected by the robot stabilization (balancing) module. However during the transition to a steady robot configuration, the leaning motions must be compensated by the camera tilt. The response time of the camera tilt actuator is faster than the robot leaning motion. As a result, the camera tilt controller compensates quickly for such robot leaning forward and backward motions, and keeps the target in the center of the image.

Many trials were performed each lasting between 20 to 60 minutes over a period of several weeks, both indoors and outdoors and under varying light conditions. The person outfit (shirt) in each trial had a single dominant color, but some also had simple patterns of one or more others colors. The environments consisted of hallways inside a building, outside paved roads, lawn, under trees, along walls, and passage through narrow spaces. Figure 13 shows two scenes in the person following. A seven minute video demonstrates the person following in indoors and outdoors [41].

Figure 13. Two snapshots of a person following trial.

We have attempted to keep the conditions of a particular walking maneuver reasonably constant. For this, we define a normal walk as that of a person walking on a straight line at a normal pace (about 1.2 m/s) in a normal environment. A normal indoors environment is hallways with the hallway lights that are turned on. A normal outdoor environment is a paved road with average lighting conditions, e.g., in the morning sun with partially clouded sky. A maneuver is defined as one that changes abruptly the course of a normal walk and/or the environment.

Maneuvers in our outdoor trials consisted of (i) Walking from sunny areas into the shades and vice versa, (ii) Suddenly changing from walking to jogging (speed from about 1.0 m/s to about 2.0 m/s), and (iii) Going from paved road to grass or uneven areas with small bumps. The following maneuvers were performed in indoors: (iv) Turning off half the hallway lights, (v) Suddenly stopping, turning on the spot and walking in opposite direction, (vi) Jumping (speed of about 3.0 m/s) one side at a 90 degree angle from the current path and continuing the walk in this direction, and (vii) going through a narrow passage about twice the width of the robot and a length of about 8 meters,

A video shows samples of the above maneuvers [41]. The main failures were in maneuver (i) due to sudden and substantial change in outdoor lighting going from sun to shade, where the system did not have enough time to quickly determine the new intensity and adjust camera gain and shutter speed.

On the other hand, lighting changes inside the building were easily compensated for. The other noticeable failure was due to jumping to a side in maneuver (vi), which the system lost the target due to inability to acquire and process sufficient number of frames. The overall success rate, i.e., ratio of total number of successful maneuvers to total number of maneuvers was 91%.

REFERENCES

[1] M. Tarokh, and P. Ferrari, "Robotic person following using fuzzy logic and image segmentation," *J. Robotic Systems*, vol. 20, No. 9, pp. 557-568, 2003.

[2] C. Schlegel, J. Illmann, H. Jaberg, M. Schuster, and R. Worz, "Integrating vision based behaviors with an autonomous robot," *J. Computer Vision Research,* Videre, 1 (4), pp. 32-60, 2000.

[3] H. Kwon, Y. Yoon, J. B. Park and A. C. Kak, "Person tracking with a mobile robot using two uncalibrated independently moving cameras," Proc. IEEE Int. Conf. Robotics Automation, 2005.

[4] T. Yoshimi, M. Nishiyama, T. Sonoura, H. Nakamoto, S. Tokura, H. Sato, F. Ozaki, N. Matsuhira and H. Mizoguchi, "Development of a person following robot with vision based target detection," Proc. IEEE/RSJ Int. Conf. on Intelligent Robots and Systems, pp. 5286-5291, Beijing, China, 2006.

[5] M. Chueh, Y.L. W. Au Yeug, K-P Calvin Lei and S. S. Joshi, "Following controller for autonomous mobile robots using behavioral cues," *IEEE Trans. Industrial Electronics,* vol. 55, no. 8, pp. 3124-3132, 2008.

[6] M. Tarokh and J. Kuo, "Vision based person tracking and following in unstructured environments," in Mechatronics and Machine Vision in Practice, pp. 99-108, Springer, ISBN 978-3-540-74026-1, 2007.

[7] S. Srinivasan, and R. Chellappa, "Image stabilization and mosaicking using overlapped basis optical flow field," Proc. IEEE Int. Conf. Image Processing, 1997.

[8] G. Chivilo, F. Mezzaro, A. Sgorbissa and R. Zaccaria, "Follow the leader behavior through optical flow minimization," Proc. IEEE/RSJ Int. Conf. on Intelligent Robots and Systems, 2004.

[9] M. Piaggio, P. Fornaro, A. Piombo, L. Sanna, and R. Zaccario, "An optical flow person following behavior," Proc. IEEE/ISIC Joint Conference, 1998.

[10] G. Chivilo, F. Mezzaro, A. Sgorbissa, and R. Zaccaria, "The robotic attendant; how to follow a leader through visual servoing," Proc. 13th IEEE Int. Workshop on Roboti and Human Interactive Communication, pp. 479-482, Genova, Italy, 2004.

[11] D. Beymer and K. Konolige, "Tracking people from a mobile platform," Int. Joint Conf. Artificial Intelligence, 2001.

[12] Z. Chen and S. Birchfield, "Person following with a mobile robot using binocular feature-based tracking," Proc. IEEE/RSJ Int. Conf. on Intelligent Robots and Systems, pp. 815-820, 2007.

[13] Censi, A. Fusiello and V. Roberto, "Image stabilization by feature tracking," Proc. !0th Conf. Image Analysis and Processing, V pp. 665-667, Venice, Italy, 1999.

[14] G. L. Foresti and C. Micheloni, "A robust Feature tracker for active surveillance of outdoor scenes," *Electronic Letters on Computer Vision and Image Analysis*, vol. 1, no. 1, pp.21-34, 2003.

[15] I, Zoghami, O. Faugeras and R. Deriche, "Using geometric corners to build a 2d mosaic from a set of images," Proc. IEEE Conf., Vision Patter Recognition, pp.420-42, 1997.

[16] W. Zajdel, Z. Zivkovic and B. J. A. Kr¨ose, "Keeping Track of Humans: Have I Seen This Person Before?," Proceedings of the 2005 IEEE Int. Conf. Robotics and Automation, pp.2081 – 2086, 2005.

[17] M. Tarokh, and P. Merloti,, "Vision-based robotic person following under light variations and difficult walking maneuvers," *J. Field Robotics,* vol. 27, no. 4, pp. 387- 398, 2010.

[18] J. Sales, R. Marın, E. Cervera, S. Rodrıguez and J. Perez, "Multi-sensor person following in low-visibility scenarios," *Sensors,* no. 10, pp. 10953-10966, 2010.

[19] Z. Chen and S. T. Birchfield, "Person following with a mobile robot using binocular feature-based tracking," IEEE/RSJ Int. Conf. on Intelligent Robots and Systems (IROS), pp. 815-820, San Diego, California, Oct. 2007.

[20] N. Bellotto, H. Hu, "A bank of unscented Kalman filters for multimodal human perception with mobile service robots," *Int. J. Social Robotics*, vol. 2, no. 2, pp. 121–136, 2010.

[21] D. Cunado, M. S. Nixon, J. N. Carter, " Automatic extraction and description of human gait models for recognition purposes," *Computer Vision and Image Understanding*, vol. 90, no. 1, pp. 1–41, 2003.

[22] M. Tarokh and R. Shenoy, Real-time Vision-based Robotic Person Following in Fast Walking," *Int. J. Imaging and Robotics,* vol. 15, no. 2, pp. 22-36, 2015.

[23] H.F. Chen, P.N. Belhumeur, and D.W. Jacobs, "In search of illumination Invariants," Proc. IEEE Computer Society Conf. Computer Vision and Object Recognition, vol 1, pp. 1254-1261, 2000.

[24] M. Kobilarov, G. Sukhatme, J. Hyams and P. Batavia, "People tracking and following with mobile robot using an omnidirectional camera and a laser," Proc. IEEE Int. Conf. Robotics and Automation, pp. 557-562, 2006. [20]

[25] H. G. Nguyen, G. Kogut, R. Barua, A. Burmeister, N. Pezeshkian, D. Powell, N. Farrington, M. Wimmer, B. Cicchetto, C. Heng, and V. Ramirez, "A Segway based robotic transporter system," Proc. SPIE Mobile Robots XVII, Philadelphia, PA., 2004.

[26] N. Dalal, B. Triggs. Histograms of oriented gradients for human detection. In Proceedings of IEEE Computer Society Conference on Computer Vision and Pattern Recognition, IEEE, San Diego, USA, pp. 886–893, 2005.

[27] X. Wang, T. X. Han, S. Yan. An HOG-LBP human detector with partial occlusion handling. In Proceedings of IEEE 12th International Conference on Computer Vision, IEEE, Kyoto, Japan, pp. 32–39, 2009.

[28] S. Tang, M. Andriluka, B. Schiele. Detection and tracking of occluded people. In Proceedings of British Machine Vision Conference, BMVC, Guildford, UK, pp. 9.1–9.11, 2012.

[29] L. Spinello, K. O. Arras. People detection in RGB-D data. In Proceedings of IEEE/RSJ International Conference on Intelligent Robots and Systems, IEEE, San Francisco, USA, pp. 3838–3843, 2011.

[30] M. Munaro, F. Basso, E. Menegatti. Tracking people within groups with RGB-D data. In Proceedings of IEEE/RSJ International Conference on Intelligent Robots and Systems, IEEE, Algarve, Portugal, pp. 2101–2107, 2012.

[31] J. Salas, C. Tomasi. People detection using color and depth images. In Proceedings of the 3rd Mexican Conference on Pattern Recognition, Springer-Verlag, Berlin, Heidelberg, Germany, pp. 127–135, 2011.

[32] Ess, B. Leibe, K. Schindler, L. V. Cool. Object detection and tracking for autonomous navigation in dynamic environments. *International Journal of Robotics Research,* vol. 29, no. 14, pp. 1707–1725, 2010.

[33] Ess, B. Leibe, K. Schindler, L. V. Cool. Object detection and tracking for autonomous navigation in dynamic environments. *International Journal of Robotics Research*, vol. 29, no. 14, pp. 1707–1725, 2010.

[34] H. Zhao, Y. Chen, X. Shao, K. Katabira, and R. Shibasaki, "Monitoring a populated environment using single-row laser range scanners from a mobile platform," in IEEE International Conference on Robotics and Automation (ICRA), pp. 4739–4745. Roma, Italy, April 2007.

[35] K. O. Arras, Ó. Martínez Mozos, and W. Burgard, "Using boosted features for the detection of people in 2d range data," in IEEE International Conference on Robotics and Automation (ICRA), Roma, Italy, pp. 3402–3407, 2007.

[36] Carballo, A. Ohya and S. Yuta, "People detection using range and intensity data from multi-layered laser range finders," IEEE/RSJ Int. Conf. on Intelligent Robots and Systems, pp.5849-5854, Oct. 2010.

[37] D. G. Lowe, "Distinctive image features from scale-invariant key points," *Int. J. Computer Vision,* vol. 60, vol. 2, pp. 91–110, 2004.

[38] J. Satake, M. Chiba and J. Miura " Visual person Identification using a distance- dependent appearance model for a person following robot," *Int. J. Automation and Computing*, vol. 10, no. 5, pp. 438-446, 2013.

[39] H. Bay, An. Ess, T. Tuytelaars and L. Van Gool, "Speeded up robust feature," *Computer Vision and Image Understanding, Elsevier*, vol. 110, pp. 346–359, 2008.

[40] B-F. Wu, C-L. Jen; W-F. Li; T-Y. Tsou; P-Y. Tseng and K-T. Hsiao, "RGB-D sensor based SLAM and human tracking with Bayesian framework for wheelchair robots", Int. Conf. Advanced Robotics and Intelligent Systems, pp. 110-115, 2013.

[41] http://www-rohan.sdsu.edu/~tarokh/lab/research-arpf.htm.

In: Robotics: New Research
Editor: Daniel G. Fisher
ISBN: 978-1-63485-967-7

Chapter 3

PORTRAIT VISION FUSION UNDER ROBOT VISION SYSTEMS

Li-Hong Juang[1,*], Ming-Ni Wu[2] and Feng-Mao Tsou[2]
[1]Engineering Research Center of Fujian Provincial Academics for Industrially Intelligent Technology and System, Engineering College, HuaQiao University, Fujian, P. R. China
[2]Department of Information Management, National Taichung University of Technology, Taichung, Taiwan, ROC

ABSTRACT

Video communications is a common way to communicate by the interactive technology, specially uses webcam for the remote interaction and see each other's characteristics from the screen display. In this research, the main goal is to augment some dynamic interactive virtual environments under robot vision system. For this achievement, a method using superimposing a segmented human portrait on a panoramic background is proposed, then the limb interactive element is added into these videos involved with a dynamic portrait segmentation method meanwhile uses Kinect device to extract a portrait for amendment, finally acquires a full of portrait information. Because the face is the most important identification region in one portrait, then the head skeleton tracking method is also used to strengthen the repair for its head segmentation, further uses the edge transparent processing to synthesis

* Corresponding author email: puuan.juang@msa.hinet.net

them into the video, therefore users can verbally and physically communicate through these video interactive ways to show more much vibrant.

Keywords: Kinect, dynamic portrait segmentation, skeletal tracking, edge transparent processing

1. INTRODUCTION

Human interaction is an important process of building relationships, comparing with the traditional telephone communications, a fast development in network and more polarization of intelligent mobile devices, the interaction not only makes people face to face but also is for long distance interaction. In recent years, the mobile devices can also use social sites (such as Facebook, and Google + and the wretch) and multimedia communication such as pictures or text to make the interactive process more pixel-rich, now has become the major means of communication for the youth groups, but it also indirectly led to the interaction is confined to the tiny device interface, therefore bow-in-response word was born. In the interactive multimedia, the video is also a common way to communicate, using network cameras let the different people around the world can see each other's screen characters and interact with each other in real time. Currently the intelligent network cameras can also be equipped with the mobile devices, video is no longer restricted by a fixed location.

In addition to the foregoing, interactive multimedia outlets can also be members of sports interaction, and in the modern equipment, it can use the somatosensory system to do limbs simulation for indoor interactive features. For 3D depth detection [1-4], because Kinect prices compared to other 3D detection equipment is low and has features to track human skeleton and the prospects for capturing portraits, it was chosen for this study. Kinect is mostly used for application on the skeleton tracking, but its dynamic portraits segmentation generally is not accurate, so for this reason, it will be segmented it by a new method. In this research, a method for superimposing a segmented human portrait on a panoramic background is proposed. The main goal is to augment dynamic interactive virtual environments.

These researchers [3] presented that the merge of color information and depth information will optimize the image segmentation by comparing with these various depth-capturing equipment. Meanwhile the researchers segment these scenes with a depth-of-field shallow. First, they cut the whole area into the foreground, the background and the undefined areas by using "trimap" classification method after watershed segmentation. Then, the undefined areas shall be processed by the "Alpha matte" method. Finally, they found that after these procedures, the captured images are more complete. From all the merits of above these surveys, here put forward the method that it can compare the uncertain areas by the color information and the depth information. Through this method, the portrait will be more complete by complementing the two information in the process of segmentation.

The proposed method is based on segmenting the human body using data from the Kinect sensor under robot vision system. The segmented human figure is obtained by means of a relative segmentation method. The research purpose is to amend the dynamic portrait segmentation from Kinect [5-6] and act the relative news from network camera for their combination, and uses background removal method [7-9] to let the different location people can enter into the same background scene, then they seem like together each other on side from picture, therefore Kinect user can follow this smart type action device holder who is in travel and do limb exchange to reached far interaction.

2. The Procedure of the Image Processing Technique

2.1. Somatosensory Technology

Somatosensory technology is that people within the detection range of the device can directly control limb movements without the use of complex operations. This technology based on somatosensory modalities and principles for its difference can be divided into three broad categories [10]: the inertial sensor, the light sensor and the combination sensor respectively. The optical sensor is the use of lasers for sending and receiving to obtain the human body images as well as the depth information in terms of strength and angle. The combination sensor is the first two categories of combination of three-axis gyroscope and infra-red sensors which can more accurately detect the wrist rotation movement.

The Kinect device equipment used in this study was developed by Microsoft, and was primarily to XBOX360 interactive game products for Windows version and commercial applications. Kinect usage has the light sensing technology and support the portrait identification and voice recognition features, and its video aspects is composed by the color camera and the infra-red launching players and infra-red receiving players, the infra-red launching players will be ray shot average projection to measure space and again through the infra-red receiving players receive each point location, further by the internal operation forms its depth images, finally through tracking the prospects of portrait depth operation from twenty human skeleton location. Because Kinect is the design for using at the body movements, it means to detect the body which is preferring based on the speed of the operation and poor accuracy [11].

2.2. Image Segmentation

Image Segmentation is the scene of calculus and an assortment of features such as color or texture to distinguish each object in response to users' needs. There are several classification methods, in this research a portrait is usually located in the foreground of a scene [9]. Segmenting the foreground image is a complex issue, some methods solve this problem as the prospects segmentation for a low depth field image [12], or use the triangulation method to detect the image of human limb for segmenting [9] and so on. For segmenting video images, it is needed to operate quickly, this research uses a dynamic portrait, a simplified method based on Kinect's depth image using trimap concept for further analysis, and use the skeleton tracking to repair the head for furthermore segmenting. The concept is a trimap categorized as the foreground, the background and the uncertainty areas, as long as the uncertainties region can be analyzed, it will be able to fully separate foreground and background images [13], which is our focusing study.

2.3. Interactive Video

Since 2005, interactive video is followed up because of the expansion of network bandwidth, internet users increasing and more using the video talking. Interactive video type can be divided into the custom interactive video, the exploring and the dialogue online interactive video [14]. The custom interactive video means that a user can view and edit with interactive components such as buttons, depending on user preferences. The exploring interactive video allows users to see through a space from multiple angles of an object, as an art show on the network can be able to rotate through multiple perspectives and carefully appreciates art works. The dialogue interactive video is now commonly used for the remote real-time video communication, using network cameras and other equipment for one to one or one to many video conversations. The applications like interactive video game for the muscles of lower limb rehabilitation training for the elderly [15] and the doctor conveys instructions to perform the rehabilitation campaign at home through the video. However these interactions are user interaction with virtual objects, therefore here proposes a new method for the multi-user offsite for physical interaction, online groups and participants of blends interactive video, then let the interactive video more real exchanges.

3. System Design and Schema

The objective of this system is to enable users for the different physical interaction between the two parties, share each other's space environment, uses dynamic portraits segmentation processing, and reduces the fusion violation produced by portrait and video. This section will describe the overall system architecture and the approach of each device. Taking Figure 1 the system process for cases, in this experiment, users are for User A and User B two parties respectively. User A, the end of part is Kinect portraits sense equipment for the produce of data, by this research of method, after computer operation processing produces dynamic full portrait segmentation. In User b, the end of part produces the simple of network depending on news, and User A, synchronize with B. This study does not take into account issues such as network traffic, mainly use User A for portrait and video fusion.

3.1. Acquiring Data

In this study, Microsoft Kinect Windows SDK v1.5 is used for acquiring somatosensory data, as Figure 2 shows, these data are for the color information, the depth and the twenty skeleton tracking points. In the depth information, Kinect device has its own searching portrait feature information, however the depth information is detected by infra-red 3D depth sensor, because infra-red is easy to has some gloss of refraction property and black of things by interference, led depth information being not completely, taking Figure 2 (d) for explanation, the direct depth portrait segmenting will let portrait segmenting being not completely, especially head is seriously, therefore next it will use the method settlement this problem.

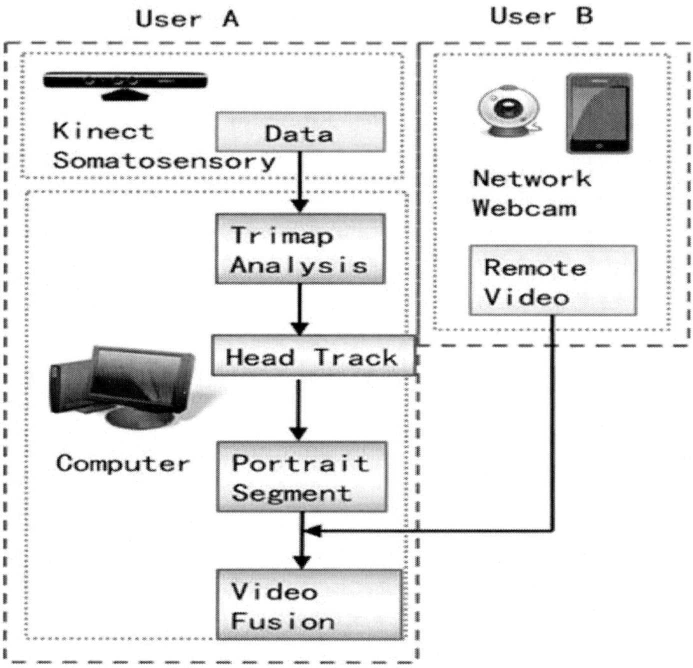

Figure 1. The system processing flowchart.

(a)

Figure 2. (Continued)

(b)

img_3_5.txt - 记事本

文件(F) 编辑(E) 格式(O) 查看(V) 帮助(H)

neck X=287, Y=152
neck X=303, Y=210
left shoulder X=245, Y=256
right shoulder X=358, Y=248
left elbow X=226, Y=318
right elbow X=395, Y=318
left wrist X=174, Y=271
right wrist X=376, Y=253
left palm X=154, Y=236
right palm X=369, Y=236
chine X=309, Y=278
middle hip X=309, Y=290
left hip X=286, Y=318
right hip X=337, Y=316
left knee X=286, Y=422
right knee X=350, Y=437
left ankle X=284, Y=575
right ankle X=363, Y=563
left thenar X=282, Y=606
right thenar X=367, Y=595

(c)

(d)

Figure 2. Somatosensory technology data from Kinect (a) color information (b) depth information (c) twenty skeletal tracking points data information (d) tracking human image segmentation.

3.2. Image Analysis and Head Tracking

In the part of image analysis, Kinect is used for face information which will be found in-depth information by trimap classification for the foreground, the background, and the uncertain areas. As shown in Figure 3, the foreground for the needs of the studied portrait is in red, the background is outside the portrait scene indicated in green and the uncertain areas is likely to not be classified the areas of foreground or background presented in white. Here also has the depth information cannot detect or correction within the error range classified as the uncertain areas. For the portrait information, it is found the damage of portrait head is in the most serious. Face is the main information identifying a person, it must be maintained its integrity, so that the special tracking for repairing the skeleton of head and neck is made.

3.3. Dynamic Portrait Segmentation

This study focuses on 640 X 480 pixels and 30 frames of color and depth information per second and designs the image processing system using Visual C # 2010 Express, as shown in the flowchart of Figure 4 in this phase. To make face more dense depth portraits image in color images, it is made adjustments based on the experimental device, for example, upward three pixels displacement, and offset one pixel to the right. Because there are many gap between the depths disparity resulting from the correction, so it expands the known 3 X 3 portrait depth image to fill up spaces between two depths.

Figure 3. The trimap image analysis.

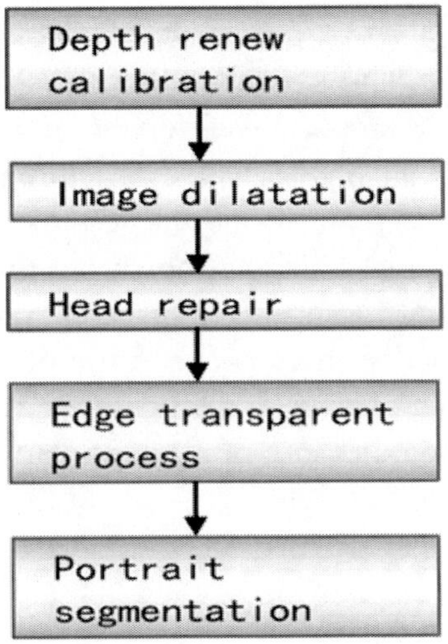

Figure 4. The procedure for dynamic portrait segmentation.

Next it is to repair the more serious damaged head, from our experience, the portrait will be included in the scope of foreground and uncertain area, and it took advantage of the head for its skeleton tracking. A portrait in the detection range of uncertainty will come into the completed known portraits of the repaired head. The detection range as Equation (1) and (2) are the followings：

$$\begin{aligned}&\mathrm{Distance}^{\mathrm{skeleton}}=\overline{\mathrm{Head}_{\mathrm{m}},\mathrm{Neck}_{\mathrm{n}}}\\&\mathrm{Distance}_{\mathrm{i}}^{\mathrm{pixel}}=\overline{\mathrm{Head}_{\mathrm{m}},\mathrm{Pixel}_{\mathrm{i}}}\\&\mathrm{T}=\mathrm{Distance}^{\mathrm{skeleton}}\times 0.6\end{aligned} \quad (1)$$

$$\begin{aligned}&\mathrm{if}((Distance_i^{pixel} < \mathrm{T})\\&\mathrm{and}\ (Pixel_i \in Unsure^{region}))\\&\mathrm{then}(Sure_i^{region}=\mathrm{true})\end{aligned} \quad (2)$$

In Equation (1), $Distance^{skeleton}$ called skeleton distance representatives for the head skeleton points $Head_m$ to the neck skeleton points $Neck_n$ of the shortest distance, and $Distance_i^{pixel}$ for the current pixel distance is by the head skeleton points $Head_m$ to the current operation pixel points $Pixel_i$ of the shortest distance. Here m, n and i are for each pixel location, and T for the detection threshold of head area is 0.6 times the skeleton distance, here times is the multiples for better measurement results under many experiments. In Equation (2), $Unsure^{region}$ representatives for the uncertain regions, $Sure_i^{region}$ represents the known portrait pixels, while the current pixel distance is less than T, and the current pixels are in the uncertain regions, here classify the pixels as the part of head and add to a known portraits region. Finally, it is moved on the known portrait for its edge transparent processing, mainly it will let the portrait synthesis can be more better fusion of video vision and reduce the cutting sawtooth on edge. However, before the transparent processing, first it must be for edge detection, because of series stream pixel information being real-time processing, it can not detect for unready pixels, therefore the left, right and top location of portrait are for edge detection. it needs to set the directional decreasing transparency matrix as Equation (3):

$$
\begin{aligned}
\mathrm{Alpha}^{\mathrm{left}} &= \begin{bmatrix} 90 & c0 & e0 \\ 90 & c0 & e0 \\ 90 & c0 & e0 \end{bmatrix} \\
\mathrm{Alpha}^{\mathrm{right}} &= \begin{bmatrix} e0 & c0 & 90 \\ e0 & c0 & 90 \\ e0 & c0 & 90 \end{bmatrix} \\
\mathrm{Alpha}^{\mathrm{top}} &= \begin{bmatrix} 90 & 90 & 90 \\ c0 & c0 & c0 \\ e0 & e0 & e0 \end{bmatrix}
\end{aligned} \tag{3}
$$

where $Alpha^{left}$, $Alpha^{right}$ and $Alpha^{top}$ are for the left, right and top decreasing transparency matrix respectively. Their value will be hexadecimal value, the value of FF is for fully opaque and the value of 00 is for fully transparent. These matrices will separately apply for the respective directions of the edge, and the edge detection equations as follows:

$$
\begin{aligned}
&\mathrm{if}((\mathrm{Pixel}_{(i\text{-}6)\sim(i\text{-}1)} \in \mathrm{Orther}^{\mathrm{region}})) \\
&\mathrm{and}\ (\mathrm{Pixel}_i \in \mathrm{Sure}^{\mathrm{region}})) \\
&\mathrm{then}(\ \mathrm{Alpha}^{\mathrm{left}}(\mathrm{Pixel}_i^{\mathrm{alpha}}))
\end{aligned} \tag{4}
$$

$$
\begin{aligned}
&\mathrm{if}((\mathrm{Pixel}_{((i\text{-}5)\sim i)} \in \mathrm{Orther}^{\mathrm{region}})) \\
&\mathrm{and}\ (\mathrm{Pixel}_{i\text{-}6} \in \mathrm{Sure}^{\mathrm{region}})) \\
&\mathrm{then}(\ \mathrm{Alpha}^{\mathrm{right}}(\mathrm{Pixel}_{i\text{-}6}^{\mathrm{alpha}}))
\end{aligned} \tag{5}
$$

$$\begin{aligned}&\text{if}((\text{Pixel}_{i-(\text{width}\times(1\sim5))}\in\text{Orther}^{\text{region}}))\\&\text{and }(\text{Pixel}_i\in\text{Sure}^{\text{region}}))\\&\text{then}(\text{ Alpha}^{\text{top}}(\text{Pixel}_i^{\text{alpha}}))\end{aligned}\qquad(6)$$

where $Orther^{region}$ is for the non-portrait zone. Equation (4) is for the left edge detection, when the operation goes through five continuous 5 non-portrait pixels, and the current pixel is located in the portrait region, then it is considered as the left edge and use its pixel transparency $Pixel_i^{alpha}$ for a center, meanwhile use $Alpha^{left}$ for the edge transparent processing. Equation (5) is especially for the right edge detection and same process as Equation (4) but use $Alpha^{right}$ for the right edge transparent processing. Equation (6) is for the top edge detection, here *width* is for the data width of color image, as same process as above, but use $Alpha^{top}$ for the top edge transparency.

Now it takes Figure 5 for the example, the red point on the head in Figure 5 (a) is for the tracking point of head skeleton, Figure 5 (b) is for Kinect internal automatically ripping portrait image, here can see some parallax correction errors and head is incomplete problem, and Figure 5 (c) is for our dynamic portrait segmenting result, our method obviously improve Kinect portrait segmenting problem and made more full of portrait image. Figure 6 is the flowchart for our actual video fusion processes. Figure 7 is the final fusion result.

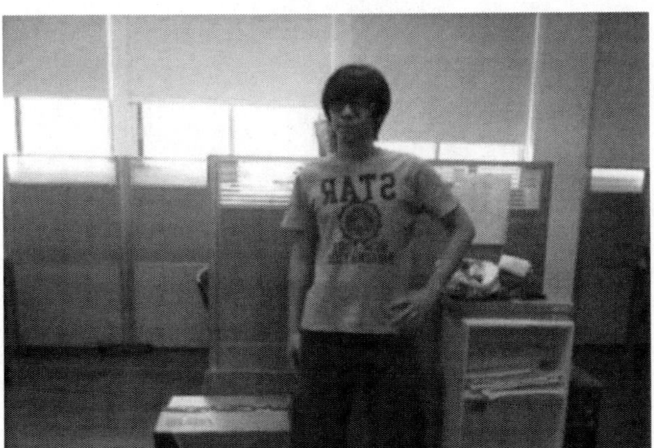

(a)

Figure 5. (Continued)

(b)

(c)

Figure 5. The results of dynamic portrait segmentation (a) Kinect head skeletal tracking (b) Kinect portrait interior detection (c) final portrait segmentation.

4. EXPERIMENT TEST

Because of the device limitations and the instruction cycle considerations, in this study, in order to simplify operation complexity for segmenting and video image fusion by thirty frames per second, here use CPU: I7-3770

3.40GHz; RAM:12GB; operating system: WIN7 64-bit; webcam devices: Logitech Pro 9000 for the computer equipment specifications.

Now it takes Figure 5 for the example, the red point on the head in Figure 5 (a) is for the tracking point of head skeleton, Figure 5 (b) is for Kinect internal automatically ripping portrait image, it can be seen some parallax correction errors and head is incomplete problem, and Figure 5 (c) is for our dynamic portrait segmenting result, our method obviously improve Kinect portrait segmenting problem and made more full of portrait image. Figure 6 is the flowchart for our actual video fusion processes. Figure 7 is the final fusion result. All the experiments show that the proposed method improves the Kinect segmentation and demonstrates that the portrait can have a relative integrity and the error rate is within 10%. Due to until now, no other similar researches and experiments can make a comparison with ours, we makes remarkable with Kinect results.

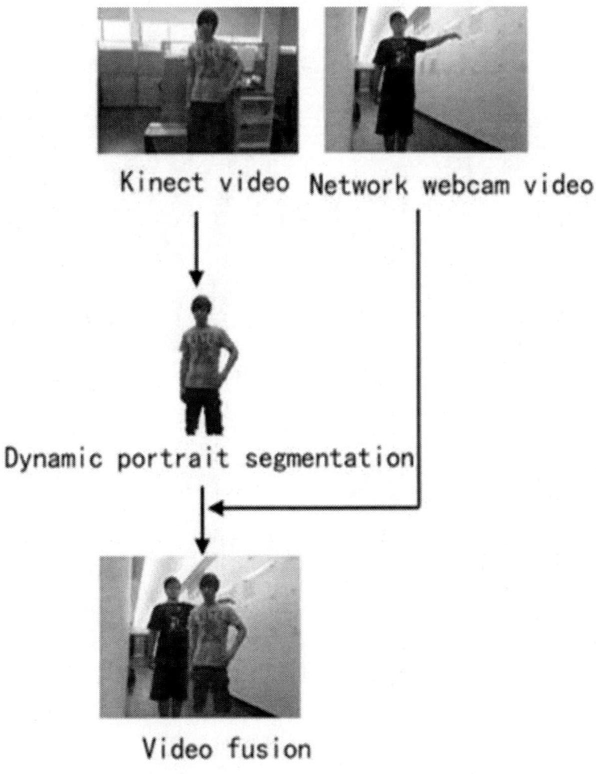

Figure 6. The system schema for interactive video fusion processing.

Figure 7. The final video fusion result.

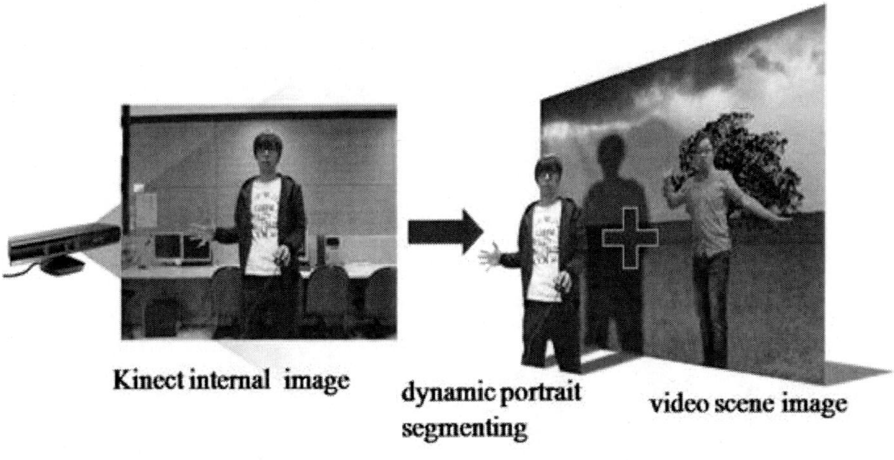

Figure 8. The video scene fusion display.

It is expected this system can reach to render a remote tour. Tourism sets up multiple networks photography in the touring scene in advance by dynamic scenery cams, users can select these views, then enter the scene in the video through the capture of dynamic portraits and set up a private interactive zones, furthermore invite some friends and families to join into this video sharing physical and verbal interaction and enhancing their relationship as Figure 8 shown for video scene fusion display. Figure 9 is the system for running dynamic portrait fusion of interactive vision in Visual c # 2010 Express screen.

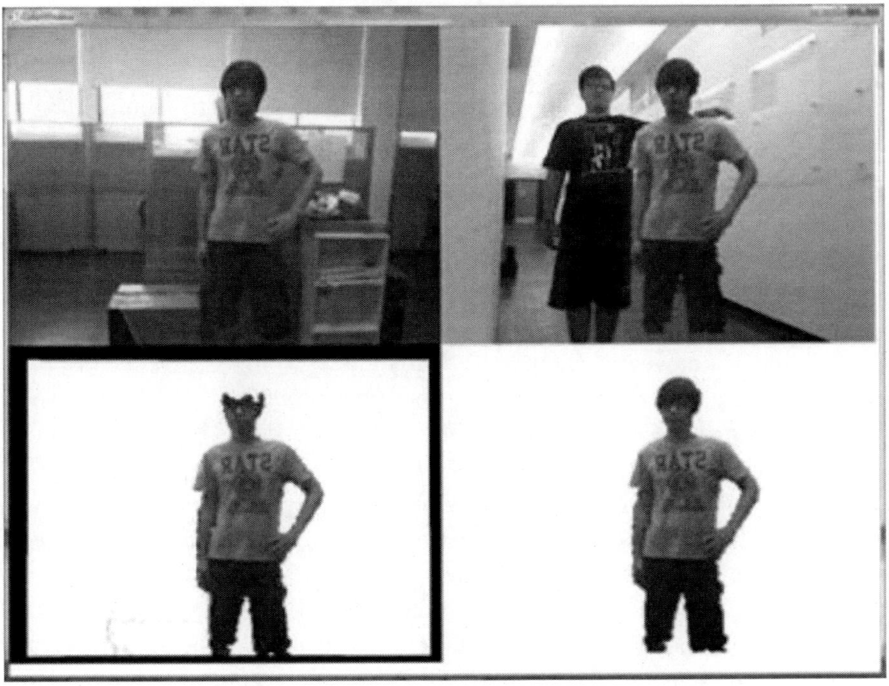

Figure 9. The serial scene for the video fusion.

CONCLUSION

This research excludes the massive computing time needed for dynamic image segmenting, the dynamic portraits segmentation can be full of expression for portraits, and particularly the head area is the most significant, segmenting good portrait by the transparent edge treatment, then it can be better integrated into the video. The main contribution is to augment dynamic interactive virtual environments by means of a relative segmentation method. The experimental results show that our method has significantly improved the Kinect segmenting problem with a complete portrait people will increase further the intention to use it for video fusion.

ACKNOWLEDGMENT

The authors deeply acknowledge the financial support from HuaQiao University, Fujian, P. R. China under the HuaQiao Scientific Research Foundation for Talents plan.

REFERENCES

[1] Asanga Ratnaweera, Saman K. Halgamuge, Harry C. Watson, "*Self-Organizing Hierarchical Particle Swarm Optimizer with Time-Varying Acceleration Coefficients,*" IEEE Transactions on Evolutionary Computation, VOL. 8, NO. 3, p. 240-255, 2004.

[2] Carme Julià, Rodrigo Moreno, Domenec Puig, Miguel Angel Garcia, "*Shape-based image segmentation through photometric stereo,*" Computer Vision and Image Understanding 115, p. 91-104, 2011.

[3] Carlo Dal Mutto, PietroZanuttigh, Guido M. Cortelazzo, "Fusion of Geometry and Color Information for Scene Segmentation," *IEEE Journal of Selected Topics in Signal Processing*, VOL. 6, NO. 5, p. 505-521, 2012.

[4] Carlo Dal Mutto, PietroZanuttigh, Guido M. Cortelazzo, "*A probabilistic approach to ToF and stereo data fusion,*" in Proc. 3DPVT, Paris, France, 2010.

[5] Chi-Yu Lee, Jin-Jang Leou, Han-Hui Hsiao, "Saliency-directed color image segmentation using modified particle swarm optimization," *Signal Processing* 92, p. 1-18, 2012.

[6] Jun-Wei Hsieh, Chi-Hung Chuang, Sin-Yu Chen, Chih-Chiang Chen, Kuo-Chin Fan, "*Segmentation of Human Body Parts Using Deformable Triangulation,*" IEEE Transactions on Systems, Man, and Cybernetics-Part A: Systems and Humans, VOL. 40, NO. 3, MAY, p. 596-610, 2010.

[7] Kittiya Khongkraphan, Pakorn Kaewtrakulpong, "*A Novel Reconstruction and Tracking of 3D-Articulated Human Body from 2D Point Correspondences of a Monocular Image Sequence,*" IEICE Transactions on Information and Systems, VOL. E94-D, NO. 5, p. 1090-1098, 2011.

[8] Luc Vincent, Pierre Soille, "*Watersheds in Digital Spaces: An Efficient Algorithm Based on Immersion Simulations,*" IEEE Transactions on

Pattern Analysis and Machine Intelligence, VOL. 13, NO. 6, p. 583-598, 1991.

[9] Wang Luo, Hongliang Li, Guanghui Liu, King NgiNgan, "*Global salient information maximization for saliency detection,*" Signal Processing: Image Communication 27, p. 238-248, 2012.

[10] China Wikipedia, *Somatosensory Technology*, available at: http://www.86wiki.com/view/5338106.htm, Accessed 7 July 2013.

[11] H. Gonzalez-Jorge, B. Riveiro, E. Vazquez-Fernandez, J. Martínez-Sánchez, P. Arias, *Metrological evaluation of Microsoft Kinect and Asus Xtion*, Measurement, 2013, pp. 1800-1806.

[12] Zhi Liu, Wei Li, Liquan Shen, Zhongmin Han, Zhaoyang Zhang, "*Automatic segmentation of focused objects from images with low depth of field,*" Pattern Recognition Letters 31, p. 572-581, 2010.

[13] O. Juan, R. Keriven, *Trimap segmentation for fast and user-friendly alpha matting, Variational, Geometric, and Level Set Methods in Computer Vision*, Lecture Notes in Computer Science Volume 3752, 2005, pp. 186-197.

[14] WIKIPEDIA The Free Encyclopedia, *Interactive video*, Available at: http://en.wikipedia.org/wiki/Interactive_video, Accessed 5 July 2013.

[15] Po-Yin Chen, Shun-Hwa Wei, Wan-Ling Hsieh, Jang-Rong Cheen, Liang-Kung Chen, Chung-Lan Kao, *Lower limb power rehabilitation (LLPR) using interactive video game for improvement of balance function in older people*, Archives of Gerontology and Geriatrics, 2012, pp. 677-682.

In: Robotics: New Research
Editor: Daniel G. Fisher
ISBN: 978-1-63485-967-7

Chapter 4

SEX RECOGNITION BASED ON ROBOT VISION SYSTEMS

Li-Hong Juang[1,*], Ming-Ni Wu[2] and Shin-An Lin[2]
[1]Engineering Research Center of Fujian Provincial Academics for Industrially Intelligent Technology and System, Engineering College, HuaQiao University, Fujian, P. R. China
[2]Department of Information Management, National Taichung University of Technology, Taichung, Taiwan, ROC

ABSTRACT

Detecting human sex from complex background, illumination variations and objects under robot vision system is very difficult but important for an adaptive information service. In this research, a preliminary design and experimental results of sex recognition will presented from the walking movements that utilize the gait-energy image (GEI) with denoised energy image (DEI) pre-processing as a support vector machine (SVM) classifier to train and extract its characteristics. The result shows that the proposed method will adopt some characteristic value and the accuracy can reach up to 100% of sex recognition rate when people are walking at a fixed angle. Meanwhile, it will be able to achieve over 80% rate within some allowed fault-tolerant angle range.

* Corresponding author email: puuan.juang@msa.hinet.net

Keywords: robot vision system, gait-energy image, denoised energy image, support vector machine, fault-tolerant angle

1. INTRODUCTION

Digital product development has made many of benefits for human life and convenient in information technology. If the recognition system in a department store or convenience store entrance can accurately distinguish between people sex and age, it will be able to provide customers with more convenient and friendly service. Think about it, if the service system can identify the customer's sex and age, then provides personal service for the different sex customer, which can reduce the time when customer searches for products, in the other hand it will also make the customer feel very intimate, and accelerate the service their guests, finally make the company's performance increasing. Sex recognition system [1] can also be developed into the security monitor system, due to today social security let people panic, more public toilets are located in the corner place, and the security let people worried. If the robot vision system can be stalled in toilet or locker room outside, when some specific strangers wander out, through recognition system will can early notify guards or the related people to handle it, so that can reduce to guard 24 hours patrol and also has a protection on the social security aspects.

Sex recognition is an important research at present. In the early, the sex studies mostly focused on face or the profile feature, but in the monitor system the face image has a low resolution or other factors lead to a lower recognition rate. Meanwhile someone being to commit crimes would deliberately wear secret and hide face feature that will result in the difficult of sex recognition. In addition, the face image is limited to help sex recognition in a long-distance monitoring, which will lead to these methods of facial feature not applied to our application. A number of key features detection of modern humans – face, race, and nutrition (and thus body shape) and age will be very difficult when somebody wears a hat, a sun glass or loose dressing. Therefore, it is found that people walking in the footsteps and the body sway exist the sex difference; male usually has the larger swing and shoulder size than female. The female has the difference in the length of hair and chest back. They will help for the sex recognition according to these characteristics. In this study, the CASIA gait dataset B [1] was used as an experimental basis, and then SVM classifier [2] used as a sex recognition source. The references [1-2] are some stock images for sex recognition system.

A lot of early studies [3-4] adapted the facial features as the sex recognition. The gait recognition can also be used for the sex recognition, Kozlowski and Cutting [5] proposed a method based on gait to distinguish the sex which used the point lights to link the joints and via observers to observe the sex, the accuracy rate can be up to 63%, although this method can be for the sex recognition but its accuracy is quite low. The other studies [6-7] also used the point lights for the sex classification, Davis and Gao [6] used an adaptive three-points model to extract the features, and their testing database contains 40 objects with 20 males and 20 females respectively, its accuracy rate can be up to 95.5%, this method's accuracy is higher than those of Kozlowski and Cutting [5] but its test data volume is small.

In 1994, Mather and Murdoch [8] proposed the method for the male and female walking with a body swing, the differences of lumbar and hip proportion and the differences of shoulder proportion, they found that the male in walking, his shoulder swing range is higher than female, but female in walking, her hip swing range is higher than male, usually male shoulder is wider than female, most female lumbar part is thinner. This research will use gait-energy image to extract the people feet feature when walking, then use the denoised energy image to remove image noise to acquire the walking physical change variance between the male and female, and then use the feature extraction as the sex recognition information. The experimental test will prove that the accuracy of our proposed method is higher than the other methods. The challenge for gait recognition is the different perspective of the marching gait which affects the correctness of the recognition. So in the paper, it will adopt the walking people in a store as object of the gait using a single camera within the allowed fault-tolerant angle range to identify its sex, therefore it will take a series of experiments to verify their accuracy and finally make some comparisons with other similar methods in the experiment tests. But because the face, race, body shape and age detections are the total different research methods and have some invalidated flaws, here it will not make a cross-comparison of these approaches.

2. THE PROCEDURE OF THE IMAGE PROCESSING TECHNIQUE

In this research, the gait-energy image (GEI) and the denoised energy image (DEI) were used as a pre-process for the gait features extraction, and then bases on the horizontal and vertical features extraction between male and female as an important characteristic reference for the sex recognition.

2.1. Procedure

In this paper, it used the people-walking gait image as the sex recognition from the original data. The procedure includes eight steps Figure 1 shows its flowchart. The first step uses CASIA gait dataset B database image for experimental data source. The second step extracts image lattice by gait sequence in the capture all shadow lattice of image. The third step uses a threshold value [9] or background subtraction [10-12] then binaries them to separate them from object and background segmentation. The fourth step makes erosion and dilation process to smooth image and removes noise as formula (1) and (2) [9], then normalize them.

$$E(x,y) = AND_{i,j=0}^{m}[f(x+i,y+j)\&T(i,j)] \quad (1)$$

$$D(x,y) = OR_{i,j=0}^{m}[f(x+i,y+j)\&T(i,j)] \quad (2)$$

where $E(x,y)$ is the erosion image, $f(x,y)$ is the binary image, $T(i,j)$ is the structure element, and $D(x,y)$ is the expanding image. m Represents the image length and width, which are equal to each other. The fifth step uses these normalized serial images for GEI [13] processing. The sixth step is the GEI image making processing to removal noise [14]. The seventh step is for features extraction. The eighth step uses the support vector machine (SVM) for sex recognition. The steps 5 through step 8 are the significant process in this research that will be explained in detail in the following section.

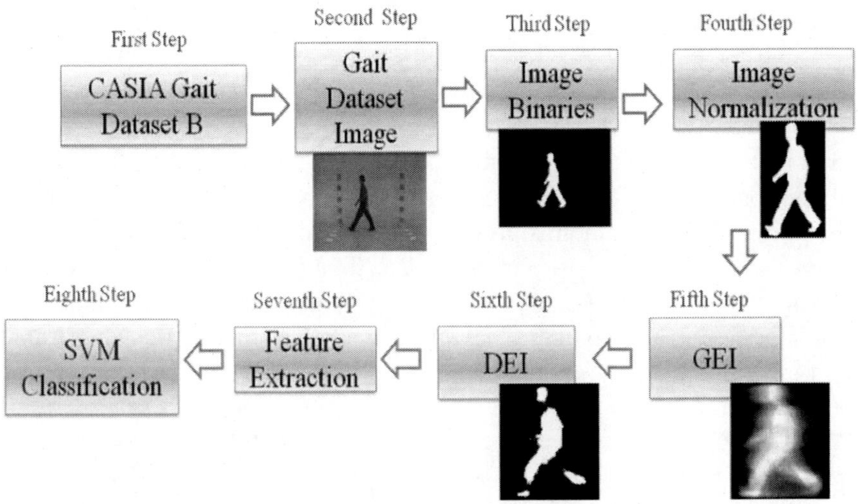

Figure 1. Flowchart for the proposed image classification procedure.

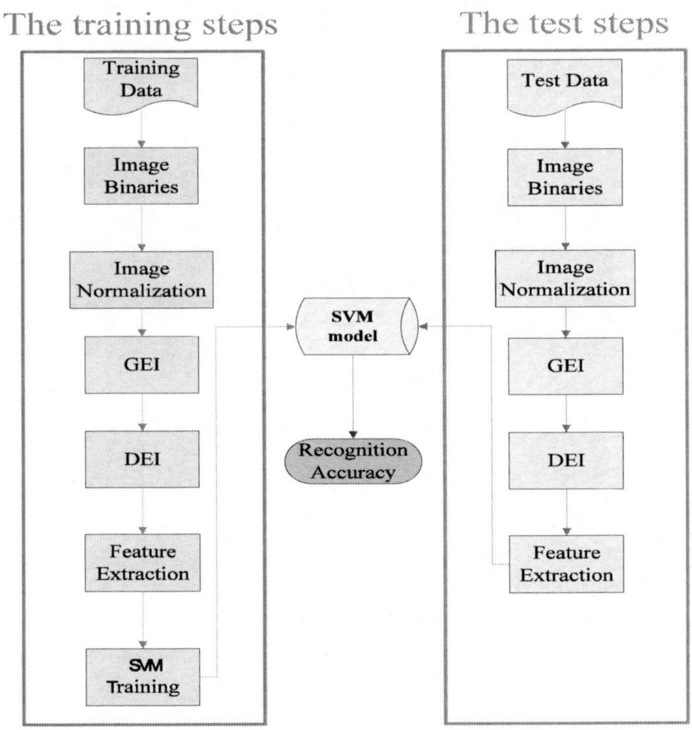

Figure 2. The training and test schema.

2.2. System Schema

Figure 2 is the system schema figure for this research method. The schema is divided into two parts respectively for training data and test data. The research method will extract the gait sequence image from movie then binaries or subtract background to acquire the prospects image, furthermore the image will be normalized and run GEI processing to acquire the serial giant variance, and then use DEI method to remove noise, finally extract the features from DEI image, then use SVM classifier to train male and female samples to acquire one best SVM model. The ultimate goal is to use the best trained model for testing and acquire the recognition.

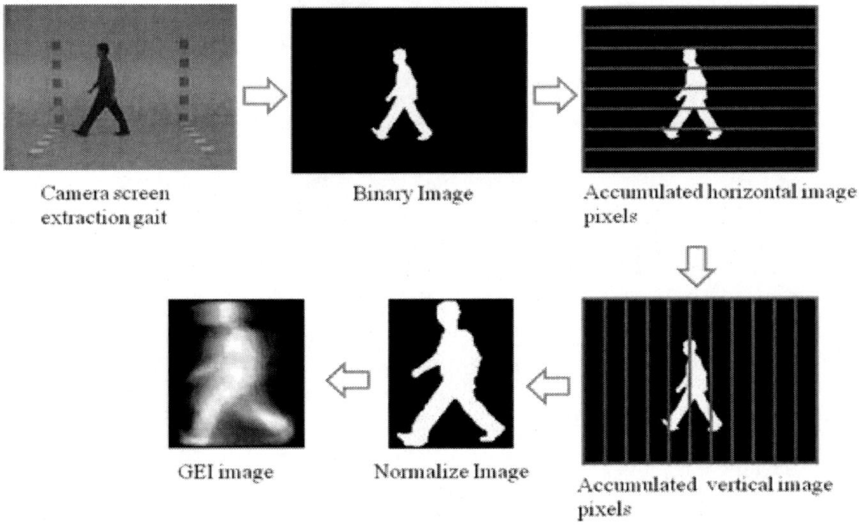

Figure 3. Gait-energy image processing.

2.3. Gait-Energy Image (GEI)

The gait is the people unique characteristics, its sex can be judged from pedestrian walking by using cameras shooting. In this research, GEI was used to calculate the gait which can dedicate the changes variance between before and after giant images. Figure 3 shows the image lattice from camera shooting, after binarization processing or background subtraction processing to acquire the foreground image and normalize it, the same serial normalized giant image lattice uses the formula (3) [14] evaluating to acquire the GEI image as shown in Figure 4.

$$G_c(x,y) = \frac{1}{N_c} \sum_{t \in Ac} B(x,y,t), \tag{3}$$

where $G_c(x,y)$ is GEI image , N_c is the total gait image serial numbers, Ac is a giant serial length and $B(x,y,t)$ representative the time image t , where x and y are the coordinates of the image pixel.

Figure 5 shows one GEI image processing example, here it is suppose d two 4 X 4 normalized image, white block is for the gait foreground image, the black block is for the background image, and gets the average image from the two images addition. If the GEI pixel value in the example image is equal to 1, it means that between the first, the second and the third image have no change. If the pixel value is 0.3 and 0.6, it means that between the first and the second and the third image have some change, so people walking characteristics can be identified.

Figure 4. The post-process image from a series of gait-energy image processing.

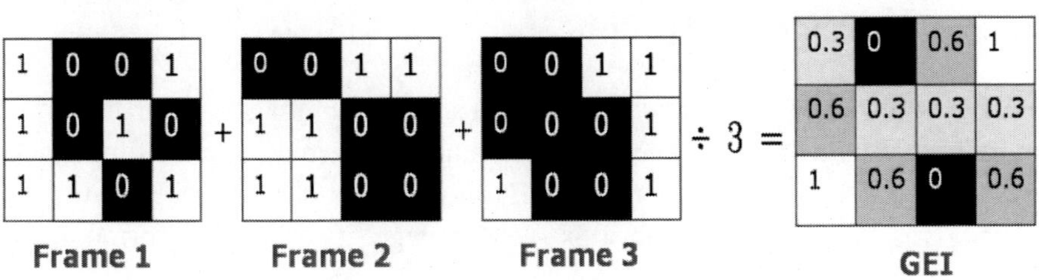

Figure 5. The gait-energy image processing example.

2.4. Denoised Energy Image (DEI)

Basically, DEI is used to remove out GEI noise, meanwhile the giant variance between male and female can be acquired. Following the formula in [14], the DEI rule is the following

$$D_c(x,y) = \begin{cases} 1, \text{if} G_c(x,y) \geq T, \\ 0, \text{otherwise}, \end{cases} \quad (4)$$

where $D_c(x,y)$ is the image after noise removal, $G_c(x,y)$ is the GEI image, and T is a threshold value. Figure 6 is the example for the GEI image making DEI processing, one 3 X 3 GEI image, the threshold value is the maximum pixel value of the GEI image (the example is 1) and multiplies 0.8 [14] to get the threshold value T, then each pixel value of GEI image compares with T, if it is greater than or equal to T, then DEI image value is equal to 1, otherwise it is equal to 0. As Figure 6 (a) shown, after the top left image of the GEI made denoising processing can become the right side DEI images, as the right side image appearance of Figure 6 (b) shown, the processed DEI image only retains the more important part of the original GEI.

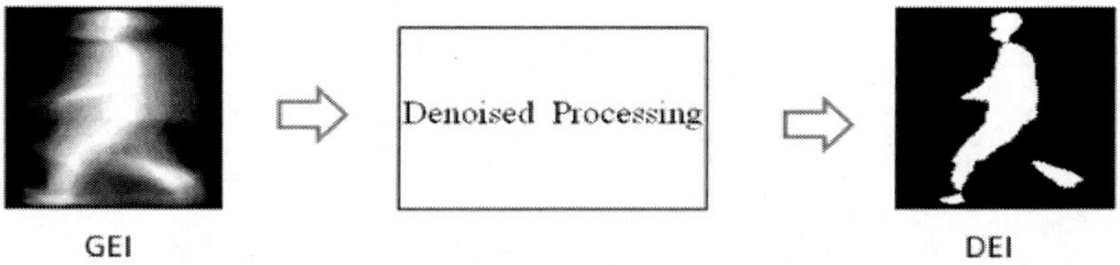

Figure 6. The denoised energy image processing example.

2.5. Characteristic Extraction

The trunk shape or hairstyle appearance of DEI images is an important feature to distinguish differences between male and female. Here we introduce how to extract the feature from DEI image, let X_i represents the ith characteristic values from the DEI extracted image, its extraction formula is as follow:

$$X_i = \sum_{k=1}^{9} c_{ik} \quad (5)$$

When $k = 1 \sim 8$, then c_{ik} as follows:

$$c_{ik} = \left\{b_{i((k-1)\times 18+l)} \middle| l = 1{\sim}18\right\} \tag{6}$$

When $k = 9$, then c_{ik} as follows:

$$c_{ik} = \left\{b_{i((k-1)\times 18+l)} \middle| l = 1{\sim}6\right\} \tag{7}$$

$$Y_j = \sum_{k=1}^{9} d_{kj} \tag{8}$$

When $k = 1 \sim 8$, then d_{kj} as follows:

$$d_{kj} = \left\{b_{((k-1)\times 18+l)j} \middle| l = 1{\sim}18\right\} \tag{9}$$

When $k = 9$, then d_{kj} as follows:

$$d_{kj} = \left\{b_{((k-1)\times 18+l)j} \middle| l = 1{\sim}6\right\} \tag{10}$$

where i is for image column, j is for the image row, k is for the block number, X_i is for the calculation feature of each column, Y_i is for the calculation feature of each row, c_{ik} is for the split block of each column, d_{kj} is for the split block of each row, and b_{ij} is for the image pixel location.

X_i Is calculated from the ith column pixel of DEI image, here DEI image size is 150 X 150 pixels, $150 \leq i \leq 1$ X_i is acquired from the 150 pixels of the ith column and the 150 pixels of the jth row in DEI. Each column and row pixels are divided into 9 blocks, c_{i1} to c_{i8} and d_{1j} to d_{8j} are responded to 18 bits, c_{i9} and c_{9j} are responded to the remaining 6 bits, and these 9 blocks numerical characteristics will be totally added for the column X_i and row value Y respectively. Therefore one DEI image will has 300 characteristic parameters for sex recognition.

2.6. Sex Recognition

Nowadays exists a lot of classifiers; however SVM was used for the classification. SVM was proposed by Vapnik in AT&T LAB team in 1995. Its main concept comes from the structural risk minimization (SRM) errors in the statistical learning theory method [15]. SVM operation is to identify an optimal hyperplane which can separate two different collections, hyperplane means a high-dimensional plane assumed used data belonging to high dimension. Taking linear core functions for its illustration, it is hoped to identify a line to separate two different sets and also want this line to enlarge the margin more and more between the two groups, so it can precisely identify the category belonging to which collection and also its calculation accuracy is better. Figure 7 shows SVM two-dimensional classification figure, if data is belonged to a high-dimensional data, then it can be projected onto the two-dimensional plane for explanation, the diagram data is divided into two groups, blue circle is as a group and the red circle is as another group, the grey part is the best line to separate the two groups. SVM classification core functions include Linear, Polynomial, and Radial Basis Function [2, 15], which formulae are as equation (11), (12), and (13) [9]

(a) Linear:

$$K(x, x_i) = x \cdot x_i \tag{11}$$

(b) Polynomial:

$$K(x, x_i) = [1 + x \cdot x_i]^q \tag{12}$$

(c) Radial Basis Function:

$$K(x, x_i) = exp\{-|x - x_i|^2 / 2\delta^2\} \tag{13}$$

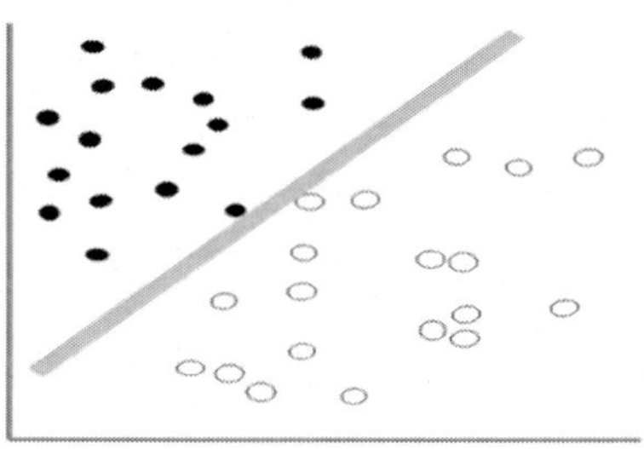

Figure 7. The two dimension support vector machine classification figure (y label is y direction location, and x label is x direction location).

Each core function has its different parameters, the parameter value also affects the classification accuracy, and the user can test multiple core functions to allow the better classification accuracy reaching the highest. In this research, Polynomial function was used for the classification recognition, which has the higher sex classification accuracy than other functions.

3. Experiment Test

This paper uses a CASI A gait dataset B image for the experiment tests which have a total of 63 test gait serial of images including 36 males and 27 females and adapt Chang and Lin [14] proposed LIBSVM software with Polynomial core functions for sex training and testing. Training image will train a SVM model for subsequent testing from the 18 males and 14 females and use two image sizes 75 X 75 and 150 X 150 for SVM model classification test. Table 1 is for image size 75 X 75 in the vertical feature extraction which accuracy is 74.19%, the horizontal feature extraction accuracy is 90.32%, and the horizontal + vertical feature extraction accuracy is 93.55%. Table 2 is for the image size 150 X 150, the vertical feature extraction accuracy is 77.42%, capture, the horizontal feature extraction accuracy is 96.77%, horizontal + vertical feature extraction accuracy is 100%, the result is reasonable due to combining the horizontal + vertical feature and using a normal image size. The experimental results show that 150 X 150 is the best image size, the vertical feature extraction accuracy is better than the vertical feature extraction, so 150 X 150 image is adapted for the horizontal and vertical feature extraction. Table 3 shows the comparison for the other four methods with our proposed method. To compare with Lee and Grimson [16] method, our proposed method has more experimental data and the accuracy is higher, then comparing with Huang and Wang [17] methods, our proposed method still has much more experimental data and higher accuracy. Furthermore with Li [18] method comparison, our recognition accuracy rate is also higher, and with Yu [19] method comparison, which used 31 male and 31 females for gait serial image test, then used SVM linear core function for training with test, feature amount will effect SVM on processing time and speed, but they used whole image as the sex features, to assume the image size 150 X 150 for an example, Yu [19] extracted 22,500 feature values, however our method only extracted 300 feature values, apparently it is more less than Yu [19] and our accuracy is higher than the previous method.

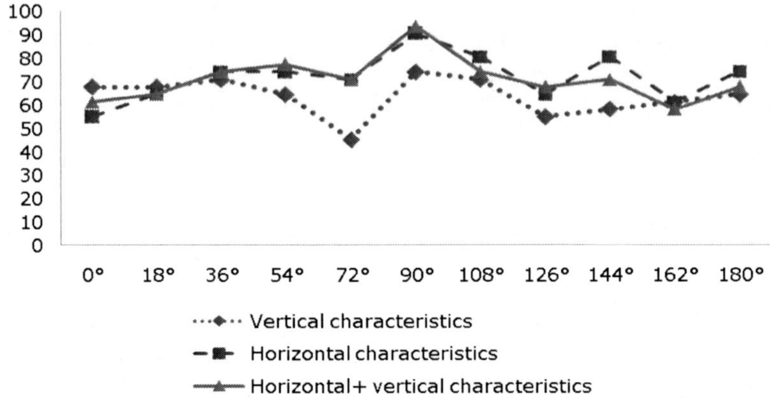

Figure 8. The feature extraction accuracy (y label) for the DEI 75 X 75 image size under the different captured angles (x label).

Table 1. The experiment results for the first image data

Image size	Characteristic extraction	CASIA Dataset	Accuracy
75 X 75	Vertical characteristics	36 males 27 females	74.19%
75 X 75	Horizontal characteristics	36 males 27 females	90.32%
75 X 75	Horizontal + vertical characteristics	36 males 27 females	93.55%

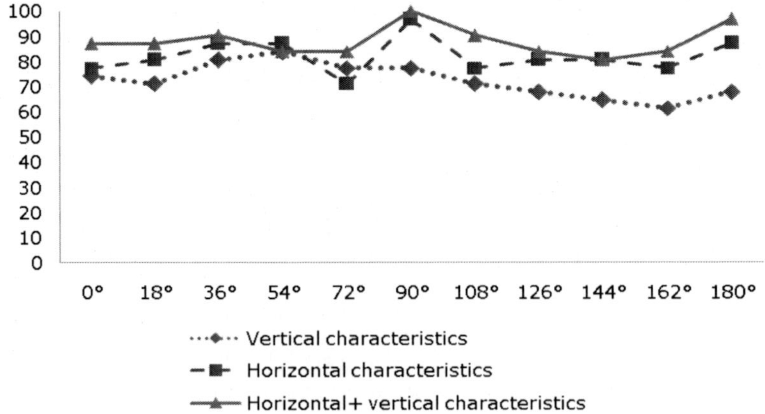

Figure 9. The feature extraction accuracy (y label) for the DEI 150 X 150 image size under the different captured angles (x label).

Now it considers the method accuracy for people walking in different captured angles. Figure 8 is the test case for the DEI 75 x 75 image size for the different captured angles, the results show that the vertical feature extraction which average accuracy is 63.64%, the horizontal feature extraction average accuracy is 72.21%, and the horizontal + vertical feature extraction average accuracy is 70.97%. Figure 9 is the test case for the DEI 150 x 150 image size for the different captured angles, the results show that the vertical feature extraction which average accuracy is 72.43%, the horizontal feature extraction average accuracy is 82.11%, and the horizontal + vertical feature extraction average accuracy is 87.98%. Therefore, it can conclude the horizontal + vertical feature extraction has the credible effect data. It also made another tests for the three kinds of camera captured angles, 90° (side), 0° (positive) and 180° (back). Table 4 lists the test using 18 males and 14 females for training data and meanwhile lists the test using 18 males and 14 females for people walking in 72°, 108°, 18° and 162°, 90° are used as the shot giant, here has a 18° fault-tolerant angle range. The result shows 80.65% and 83.87% accuracy for 72° and 108° shot angle respectively. The test shot angle 18° has 80.65% accuracy in terms of 0° (positive) shot and the test shot angle 162° has 87.1% accuracy in terms of 180° (back) shot. Due to the accuracy not achieving the idea condition, it adds more train data. Table 5 lists the training data using 20 males and 20 females and the test data using 36 males and 27 females, if 90° is for the training data, 72° and 108° are for the test data then the accuracy can be up to 84.12%. If 0° is for the training data, and 18° is for the test data, then the accuracy can be up to 82.54%. If 180° is for the training data, and 162° is for the test data, then the accuracy can be up to 87.3%.

Table 2. The experiment results for the second image data

Image size	Characteristic extraction	CASIA Dataset	Accuracy
150 X 150	Vertical characteristics	36 males 27 females	77.42%
150 X 150	Horizontal characteristics	36 males 27 females	96.77%
150 X 150	Horizontal + vertical characteristics	36 males 27 females	100%

Table 3. The experiment results for the third image data

Methodology	CASIA Dataset	Accuracy
Lee and Grimson [16]	25 males 25 females	85.0%
Huang and Wang [17]	25 males 25 females	85.0%
Li et al. [18]	31 males 31 females	93.28%
Yu, Tan, Huang, Jia and Wu [19]	31 males 31 females	95.97%
Our method	36 males 27 females	100%

Table 4. The first set data accuracy for the three different shot angles

Training data (18 males 14 females)	Test data (18 males 14 females)	Accuracy
90°	72°	80.65%
90°	108°	83.87%
0°	18°	80.65%
180°	162°	87.1%

Table 5. The second set data accuracy for the three different shot angles

Training data (20 males 20 females)	Test data (36 males 27 females)	Accuracy
90°	72°	84.12%
90°	108°	84.12%
0°	18°	82.54%
180°	162°	87.3%

CONCLUSION

In this research, it has proposed a realization of sex detection from walking movements using gait-energy image with denoised energy image pre-processing as support vector machine classifier. A lot of evaluation on the image database shows encouraging results which accuracy has reached to 100% under vertical test. And the method accuracy for people walking in three kinds of different captured angles also can be above 80%. In the future, it will upgrade the proposed method to use for a holding object condition under variant filming condition.

ACKNOWLEDGMENT

The authors deeply acknowledge the financial support from HuaQiao University, Fujian, P. R. China under the HuaQiao Scientific Research Foundation for Talents plan.

REFERENCES

[1] http://www.cbsr.ia.ac.cn/english/Gait%20Databases.asp.

[2] http://www.csie.ntu.edu.tw/~cjlin/libsvm/index.html.

[3] Golomb B. A., Lawrence D. T., and Sejnowski T. J., *Sex Net: A neural network identifies sex from human faces*, in Proc. Conf. Advances in Neural Information Processing Systems 3, 1990, pp. 572-577.

[4] Jain A. and Huang J., *Integrating independent components and linear discriminant analysis for gender classification*, in Proc. 6th IEEE Int. Conf. Automatic Face and Gesture Recognition, 2004, pp. 159-163.

[5] Kozlowski L. T. and Cutting J. E., *Recognizing the sex of a walker from a dynamic point-light display, Percpt.* Psychophys, vol. 21, pp. 575-580, 1977.

[6] Davis J. W. and Gao H., *Gender recognition from walking movements using adaptive three-mode PCA*, in Proc. Conf. Computer Vision and Pattern Recognition Workshop, Washington, DC, 2004, vol. 1, p. 9.

[7] Troje N. F., Decomposing biological motion: A framework for analysis and synthesis of human gait patterns, *Journal of Vision*, vol. 2, no. 5, pp. 371-387, 2002.

[8] Mather G. and Murdoch L., *Gender discrimination in biological motion displays based on dynamic cues*, in Proc. Biological Sciences B, 1994, pp. 273-279.

[9] Xue Z., Ming D., Song W., Wan B. and Jin S., *Infrared gait recognition based on wavelet transform and support vector machine*, Pattern Recognition, Vol. 43, 2010, pp. 2904-2910.

[10] Sarkar S., Phillips P. J., Liu Z., Vega I. R., Grother P., Bowyer K. W.,

The human ID gait challenge problem: data sets, performance, and analysis, IEEE Transactions on Pattern Analysis and Machine Intelligence, Vol. 27 (2), 2005, pp. 162-177.

[11] Huang P. S., Harris C. J., Nixon M. S., *Human gait recognition in canonical space using temporal templates*, IEEE Proceedings - Vision, Image, and Signal Processing, Vol. 146 (2), 1999, pp. 93.

[12] Wang L., Tan T., Ning H., Hu W., *Silhouette analysis-based gait recognition for human identification*, IEEE Transactions on Pattern Analysis and Machine Intelligence, Vol.25 (12), 2003, pp. 1505-1518.

[13] Lam Toby H. W., Cheung K. H. and Liu James N. K., *Gait flow image: A silhouette-based gait representation for human identification*, Pattern Recognition, Vol. 44, 2011, pp. 973-987.

[14] Chen C., Liang J., Zhao H., Hu H. and Tian J., *Frame difference energy image for gait recognition with incomplete silhouettes*, Pattern Recognition Letters, Vol. 30, 2009, pp. 977-984.

[15] Gunn S. R., *Support Vector machines for classification and regression*, Technical Report University of Southampton, 1998.

[16] Lee L. and Grimson W. E. L., *Gait analysis for recognition and classification,* in Proc. 5th IEEE Int. Conf. Automatic Face and Gesture Recognition, Washington, DC, May 2002, pp. 155-162.

[17] Huang G. and Wang Y., *Gender classification based on fusion of multi-view gait sequences*, in Proc. 8th Asian Conf. Computer Vision, 2007, pp. 462-471.

[18] Li X., Maybank S. J., Yan S., Tao D., and Xu D., *Gait components and their application to gender recognition*, IEEE Trans. Syst., Man, Cyber. C, Appl. Rev., vol. 38, no. 2, pp. 145-155, 2008.

[19] Yu S., Tan T., Huang K., Jia K. and Wu X., *A Study on Gait-Based Gender Classification*, IEEE Transactions on Image Processing, Vol. 18, 2009, pp. 1905-1910.

In: Robotics: New Research
Editor: Daniel G. Fisher

ISBN: 978-1-63485-967-7

Chapter 5

ROBOT VISION METHODOLOGY TO DETECT CRACKS

Li-Hong Juang* and Wei-Jen Chen
Engineering Research Center of Fujian Provincial Academics
for Industrially Intelligent Technology and System, Engineering College,
HuaQiao University, Fujian, P. R. China

ABSTRACT

Currently, the security of large civil engineering structures more and more people's attention, health monitoring study civil engineering at home and abroad is one of the hot topics of research. After a hundred years of research, the traditional structural health monitoring theories and methods continue to improve, and now there have been a lot to be able to use the results of the actual engineering and have been widely applied to large bridges and other structures, but there is a lot of its own defect. With the development of computer technology, people began to use computer vision to health monitoring, and gradually formed a monitoring method of mechanical vision. This paper introduces the significance and application health monitoring methods, and some of the traditional health monitoring methods are introduced. Then reviewed the development and the use of machine vision health status monitoring, and finally to crack measuring, for example, using wavelet recognition method for image recognition and analysis of fracture.

---

* Corresponding author email: puuan.juang@msa.hinet.net

Keywords: health monitoring, damage identification, machine vision, wavelet analysis, fracture identification

1. Introduction

Currently, the study of structural health monitoring method in the world are a hot topic. In 1996, control of the General Assembly on international structures scholars have proposed the establishment of three specialized working groups for structural health monitoring in Europe, Asia and the United States. Since 1997, every two years regularly convene an international agenda for structural health monitoring at Stanford University. In 2000, the American Society of Civil Engineers in the organization of the "structural health monitoring of standards," the research topic. Our country is in the mid-1990s under the auspices of "Climbing Plan" and the National Natural Science Foundation began to study the problem of structural health monitoring [1].

Overall, structural health monitoring in civil engineering is still a relatively new discipline, after a hundred years of research Despite many achievements, but many of its findings have not made use of the civil engineering go, confined to laboratory research stage [2]. Structural Health Monitoring began in the 1950s abroad, initially mainly used in aerospace and mechanical and other fields. Early health monitoring tools rely mainly on manual monitoring of experienced engineers and structural health status and trends judgment [3]. Research in the field of civil engineering is probably the early 1980s, researchers mainly use aspects of health monitoring research in bridge engineering and high-rise buildings. In the 20th century, 70-80 years, in order to monitor the safety of offshore oil platforms, people have done a study [4] use health monitoring system to monitor mining platform. Currently, structural health monitoring study abroad in civil engineering mainly in four aspects: reservoir (dam), bridges, maritime measures, high-rise building [5]. In addition, in the protection of historical heritage buildings, public measures have also been using roads.

Structural health monitoring through a hundred years of research, now have a relatively complete theoretical system. In Study sensor technology, Sun and other changes in the electrical impedance presented by PZT monitoring structural damage, the current, thereby the piezoelectric transducer made of theory has been widely used. New sensors, in 1978, Hill et al. found that the photosensitive phenomenon in Canada, and the use of legal standing create the world's first fiber grating. In 1989, the United States Meltz invented a germanium-doped fiber outer purple write technology, these studies so that the optical fiber manufacturing costs, it becomes possible to make extensive use of [6], in theory, the sensor arrangement is currently the most widely used by the Kammer [7] proposed effective independence method (EI). Theoretical study abroad in damage recognition is also very rich, in the 1960s at the University of Michigan Professor Holland America proposed genetic algorithms, such as the use of genetic algorithms thereafter Koh local search, by measuring excitations and responses to structural parameters identification, discovery load Location great impact on the recognition results [8]. In addition to genetic algorithms, wavelet transform scholars have also done a lot of research, such as the Mexican Hat Wavelet Segawa such as the use of the system input acceleration and acceleration response analysis to identify the damage caused by the parameter structure, Hear wavelet analysis theory, etc. Health Monitoring Benchmark American Society of Civil Engineers (ASCE) proposed (health monitoring benchmark problem) were studied, and the results of wavelet theory especially suitable for on-line monitoring of the conclusions [9], in 1992, the famous French Information Sciences IR ISA institution of Zhang et al. [10] proposed a wavelet neural network concepts and algorithms. With the development of computers, scholars began to try to use image processing techniques to study structural health monitoring. In 1983, American University of South Carolina's M, A, Sutton made a digital image correlation method [11], later scholars through research efforts, now photo grammar has been commercialized abroad [12]. US logistics home Satty in the 1970s proposed AHP in safety assessment methods research system, professor Zadeh proposed fuzzy comprehensive evaluation method [9] in 1965.

In China, structural health monitoring is mainly used on a number of important large-span bridges, many researchers also research-based bridge monitoring. In the study of Tsinghua University sensor arrangement problem right Qin et al. Tsing Ma Bridge in Hong Kong optimal sensor placement problem is studied, and opened up a new way better to solve the large-scale structure of optimal sensor placement problem [13]. In addition, the research and application of structural safety aspects of comprehensive evaluation and decision-making, such as Wu, GU Chong, Shen proposed and developed a medium build a machine based on the Four Dam Safety Comprehensive Appraisal Expert System, the system pattern recognition and fuzzy evaluation, inference engine through an integrated, comprehensive call for Siku, the quantitative and qualitative analysis combined to achieve dam safety status online real-time analysis and comprehensive evaluation [14]. Currently, the system has been used in our Danjiangkou Dam and LONGYANGXIA

Gutianxi dam, which provides a timely assessment of the security system for the dam [15]. Both some of the existing methods of research and practice in terms of damage detection methods, scholars opened up some new way. Eg due to bridge structural health monitoring with real-time and data volume characteristics, Guo Jian have put forward an application of two real-time test data separately wavelet decomposition and reconstruction method for the expression of signal characteristics, and defines a damage indication coefficients [16]. Some scholars have proposed their new methods, such as Professor Deng Julong issued a "gray system control problems," a paper in 1892, the creation of gray system theory [17]. In high-rise building health research monitoring methods Despite some progress, but the technology is still very immature [3]. Currently, most of the building structure to monitor the use of domestic mainly used in the construction phase [18]. Also, with some scholars such as Wang Juandeng [19] cultural heritage building health monitoring system, a number of studies.

Structural damage identification is the core of structural health monitoring, at present, structural damage identification method according to the main structure for the monitoring can be divided into the overall law and local law, by way of damage identification method can be divided into static and dynamic law. Overall method which will structure as is a system of mass, stiffness, damping and other physical parameters of the composition, and the structure of its physical parameters monitored by systematic research, the current overall method includes dynamic fingerprint analysis, model updating and system identification, neural networks, genetic algorithms and wavelet analysis method. Local law mainly dependent on the power of modern technology, the use of certain instruments of quantitative indicators to analyze the internal structure. Currently, local law mainly visual inspection method, staining, indentation method, rebound, ultrasonic pulse method, radiation method, magnetic particle method [20]. Wavelet recognition method from the traditional Fourier analysis, in order to solve the traditional Fourier analysis capabilities of local produce poor. Overseas Research on wavelet analysis earlier, in 1982 JOStromberg to construct the first close small Bo Jiji, But at the time it did not cause people's attention. 1986 Meyer constructed the first real Meyer wavelets base and proposed the idea of multi-scale analysis. Later Mallat proposed a multi-resolution analysis of the concept, and the establishment of Mallat algorithm of wavelet theory to the study of the algorithm makes multi-directional applications in many fields [21]. In the 1990s wavelet study abroad several times to form a research boom, wavelet research is in the 90s of the last century began to study, and in 1990 with the relevant papers published publicly. Our research in the wavelet to wavelet research-based, and wavelets on the use of digital image processing accounted for a substantial proportion [22]. Wavelet analysis is a way to put the data, function or operator is decomposed into different frequency components of the tool, in large structures (such as bridges) vibration condition monitoring, the wavelet analysis relative to other ways to reduce the number of sensors. After some time on the vibration point on some of the complex structure corresponding wavelet decomposition reaction of the vibrational state can be the entire structure. Thus, wavelet analysis is applied to the vibration monitor real time online.

While traditional theory of health monitoring method is perfect, and by the piezoelectric sensors, light barriers when health monitoring sensor accuracy is very high, in terms of measurement variability can achieve accuracy of hundredths of a millimeter, but there are also small measurement range, the measurement process sensor loss, poor durability [23]. In addition, traditional health monitoring system large and complex bridge monitoring system as an example to our country, in general, a detection system consists of at least 50, as many as more than 300 sensor points, whose cost accounts for the total cost of the bridge 0.5 about %, while health monitoring system installation costs can account for about 10 percent of the entire system [24]. The machine vision with a non-contact method of monitoring, adaptability, economic, easy to operate, etc., help to overcome the shortcomings of traditional monitoring methods, development potential. In addition, people cannot reach or hard to reach places, machine vision of health monitoring method with respect to the conventional method has great advantages.

The first image processing technology produced in the 1920s to produce in the newspaper industry, was just a simple image compression, but it also can greatly reduce the time to transfer the image [25]. Digital image processing technology can be applied to other areas, such as medical, aviation and other aspects of development in the 1960s thanks to the computer. Began in the 1970s, due to the development of computer technology and artificial intelligence research, people began to study the use of a computing system image analysis method, this method is similar to people to get information from the image, called machine vision [26]. Subsequently, in 1977 the Massachusetts Marr proposed computational theory of machine vision, the theory has become a guide in the 1980s in the field of machine vision research theoretical framework [27]. Mid-80s, machine vision theories flourish, digital image processing technology developed to compress the image conversion from the beginning of time image, image enhancement, image restoration, image edge enhancement, image segmentation and other techniques [28].

The scope of the current study and application of machine vision systems are very extensive. In the industrial inspection, in the 1980s, Canada Diffracto company installed a laser system for detecting auto welding and assembly is qualified, its precision can reach 0.1mm [29]. In the medical X-ray, CT detection, traffic areas for vehicle identification in the field of satellite remote sensing for environmental monitoring, Geodetic [30]. In agricultural production, machine vision technology can be used to grow crops to detect, analyze its growth status and degree of damage; can be used to identify the quality of agricultural products, by digital image processing techniques to analyze the color, surface defects, etc. to evaluate its quality; can also be used for agricultural Harvest regard, in 1991, Japan "Kubota, 'the company has successfully developed a robot for the orange harvest, the robot can automatically identify ripe oranges in the natural environment, the accuracy rate up to 75% [31]. Although the machine vision technology has been applied to many areas, but not yet widely practiced law, including both technical imperfections of reasons, more important is the machine vision system efficiency depends on the efficiency of digital image processing algorithms, and the algorithm is not old at home and abroad method to solve many practical problems, and new algorithms are still a lot of theoretical research stage [32]. In terms of the use of civil engineering, foreign countries have begun to make some achievements, but China is still in the research stage. In 1996, Scott will Photography Measurement techniques applied to soil mechanics tests, using digital image correlation method for measuring the success of a strain of soil under stress state distribution [46]. Tongji University YF Ji et al. 2008 also published a related paper, the use of photography Measurement indoors cantilever vibration test done in the external force, by measuring the displacement method, measure its natural frequency, and on a cable-stayed pedestrian bridge to do the appropriate tests, the successful measure its natural frequency [33].

Currently, the domestic health monitoring in machine vision is still in the research stage, the contents of the study to identify and measure the main crack, in recent years there have been many achievements. Zhejiang University leaves precious as et al. [34] using the self-programming on the analog cracks (length 100mm, width 3.60mm) identify measurement error is only 1.6% of its width. In addition, the use of digital image processing techniques to measure the actual number of cracks, the crack width is greater than when the comparison of the results with the data 0.4mm crack width gauge measured, the error can be controlled within 10%. Thus, digital image processing technology which has a certain accuracy, to meet the requirements of engineering practice.

Central South University Weijun et al. [35] measured the crack image processing to analyze key issues of an engineering beam and wall cracks actual measurements. This measurement requires only a digital camera and a tripod, and the results maximum error of between 3% to 16% (Note, the width of the crack is measured 0.21mm ~ 1.21mm).

Comparison of Southeast University Nie Honglin, Hu Wu-sheng [36] With each use ruler, cracks and Digital Image Processing Technology in Nanjing in the Ming Dynasty city wall crack monitoring carried out, and the results of the three methods detected by the line. Among them, the digital image processing technology by way of a digital camera and tripod free access point data acquisition, image processing by matlab7.0 programming, measure the width of the walls of cracks. The results of crack width ruler measuring the resulting digital image processing, the result crack observation instrument measurements were compared, the average difference between positive and negative ruler measuring only 0.08mm, and crack meter average difference of plus or minus 0.06mm. More single research domestic and foreign health surveillance in machine vision with respect to its contents and a wider range, and many achievements have been applied to the actual project.

Ioannis Brilakis and Lucio Soibelman [37] studies the use of machine vision surveillance methods to monitor construction, with the image processing method of the construction site transfer of picture identification, performance assessment materials. ZhiQiang Chen and Tara C. Hutchinson [38] using digital image processing techniques to satellite images of Iran Bam region 6.6 earthquake occurred in 2003 after the analysis, using a new edge detection method for earthquake damage assessment of urban structures.

2. Math Model

Wavelet transform is a method of time-frequency analysis of a very good, it has a very widely used in digital image processing, such as image compression, image enhancement, wavelet de-noising, image matching, feature extraction, edge detection. We will be using wavelet analysis crack image processing, build recognition model, which is mainly applied to the wavelet decomposition and reconstruction method.

2.1. One-Dimensional Continuous Wavelet Transform

We define the continuous wavelet transform as follows:

In the formula (1), we called mother wavelet or tolerance Xu Xiaobo, where the wavelet transform scale factor, displacement factor. In signal processing, continuous wavelet transform effect is actually the signal into a binary function elements become, it can be said that the signal in the wavelet decomposed into different frequencies, similar to the role of physics in the white Mitsubishi mirrors decomposition into different colors of light.

2.2. Discrete Wavelet Transform

Since the computer is used in digital processing mode, we need to implement the wavelet analysis in a computer requires a continuous wavelet transform discrete, in addition, discrete can also reduce and eliminate the redundancy of one-dimensional wavelet transform coefficients. Generally can scale factor or displacement factor discrete processing. In discrete displacement factor, for example, so that, in practice often take you, you can obtain the following discrete wavelet transform.

Discrete wavelet transform is shown in Figure 1.

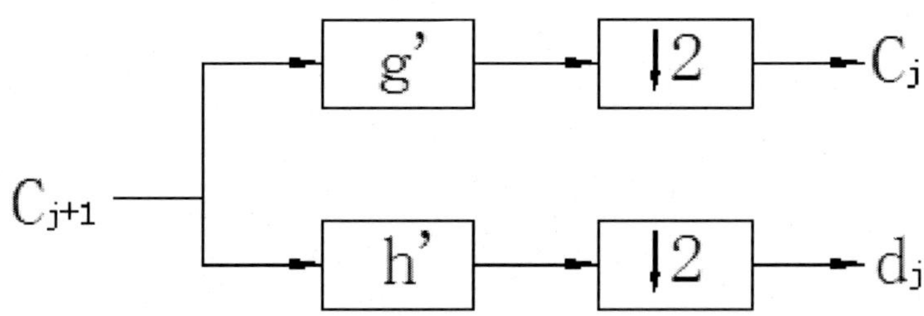

Figure 1. One-dimensional discrete wavelet.

In the above decomposition, we called the approximate sequence, known details sequences and are high-pass filtered output and low-pass filtered output, and represents downsampling operator, by discrete wavelet transform can decompose the signal into high-frequency and low-frequency part section. By wavelet decomposition we can extract the required part, reconstructed, reconstruction principle and contrary decomposition principle.

2.3. D Wavelet Transform

The difference is that the signal is a two-dimensional image scaling function, we can decompose by a one-dimensional in both directions through the line is completed, the decomposition process is shown below.

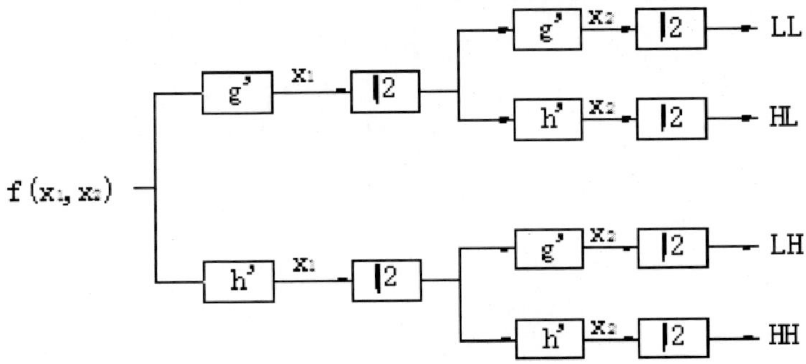

Figure 2. Dimensional wavelet decomposition.

Among them, the two-dimensional wavelet decomposition of the image twice, first for the direction, the second direction. Is defined herein as an image in the horizontal direction, the vertical direction, L represents low frequency, H represents high frequency. During the decomposition process, because we use the image of the operator second extraction, resulting in post-processing the image size becomes the original image, the image-level results can be obtained by decomposing the image expressed in the following Figure:

LL	HL
LH	HH

Figure 3. An exploded image.

Similarly, the images can be multi-level exploded, Figure 4 shows an exploded view of two crack 5. schematic exploded.

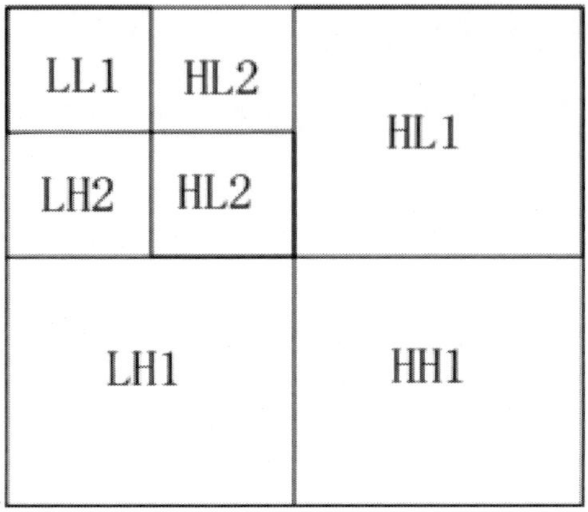

Figure 4. Image two decomposition.

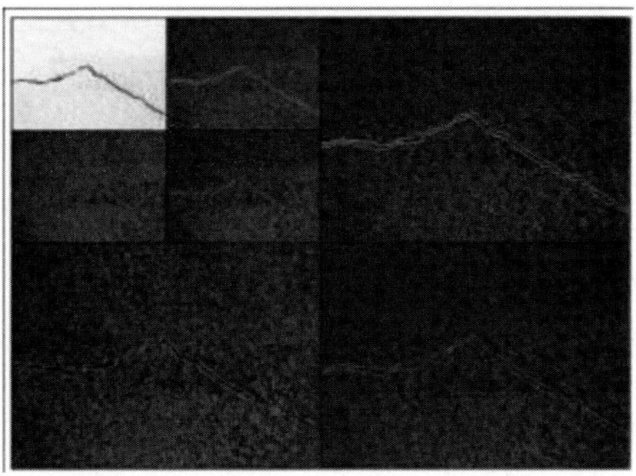

Figure 5. Image two Effectiveness.

3. Fracture Identification Based on Wavelet Analysis Model

In this section, mainly using matlab Wavelet toolbox in the original fracture decomposition, and then extract the corresponding part of the analysis. It also needs to be extracted from the original image or an image some image enhancement, image denoising, and some morphological processing, the entire workflow general procedure shown in Figure 6, a method which uses a specific image corresponding to the different, wherein image pre-processing and post-processing is not necessary, in this section will be on some of the important steps below explain in detail.

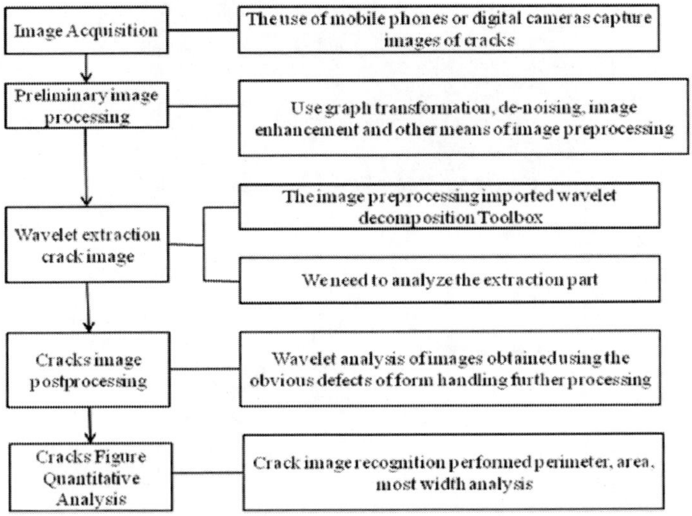

Figure 6. Schematic model.

4. The Initial Image Processing

Original image on your phone or digital camera acquisition, since the image is true color image capture, processing in matlab served to convert it to grayscale images, for on the other hand, since the image itself contains a lot of noise, will wavelet recognition effect have some impact, we can first of the original image denoising results shown in Figure 7, 8, 9, 10.

4.1. Wavelet Decomposition

Preprocessed image, we can use some of the features in matlab Wavelet toolbox crack image decomposition in matlab wavelet toolbox can choose a variety of wavelet decomposition several times, its principle, see Section 4, and then extract desired part. In this experiment exploded through practice taking first image, that diagram (5.4) in the marquee part of the process 11 and 12.

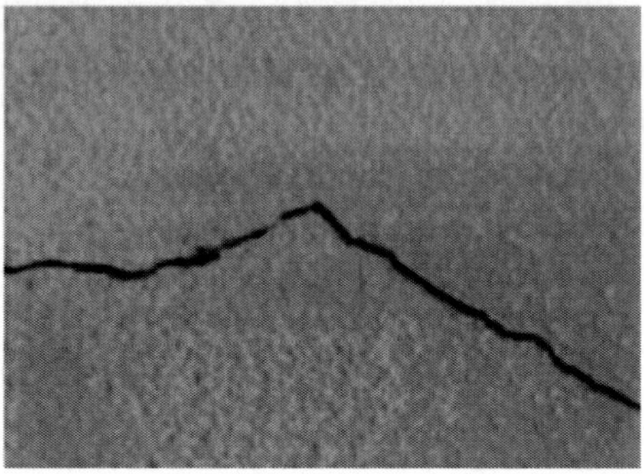

Figure 7. Original image.

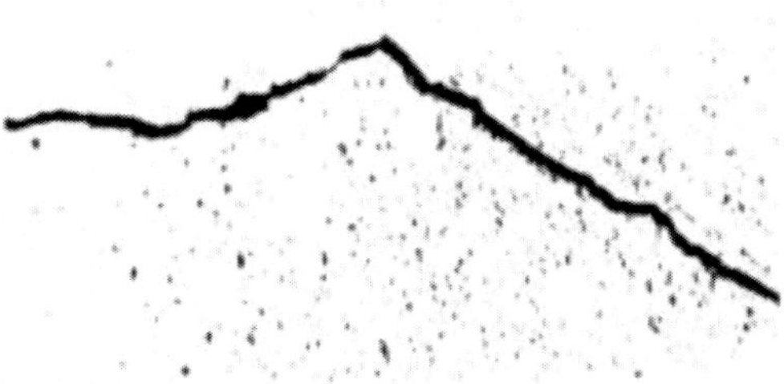

Figure 8. Binary picture.

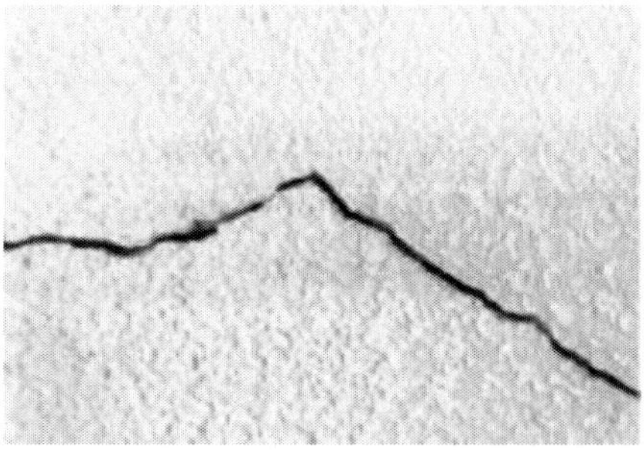

Figure 9. Wavelet denoising image.

Figure 10. Binary wavelet denoising.

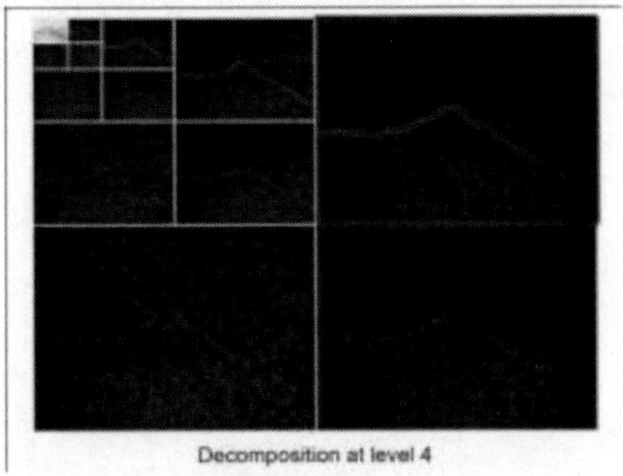

Figure 11. Wavelet decomposition toolbox.

Figure 12. Extracts part of the decomposition.

4.2. Image Post Processing

Wavelet extracted from the image by image because it is gray, in order to facilitate quantitative analysis, can make some binarization, negated the operation, some of the cracks in the wavelet identified other issues there is not a continuous image can also be some form treatment, treatment results in Figure 13, 14, 16.

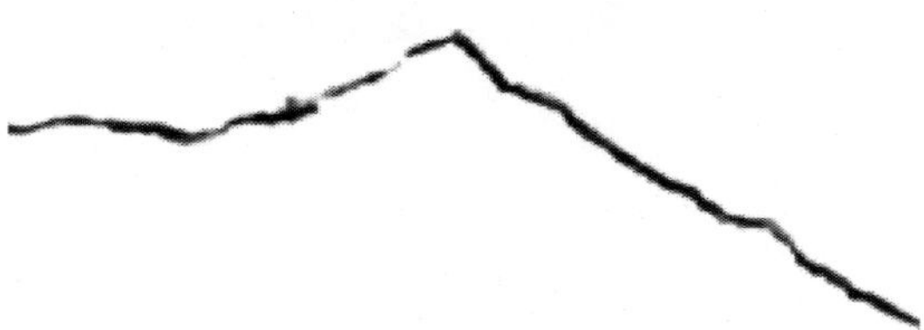

Figure 13. On Wavelet image.

Figure 14. Binary image.

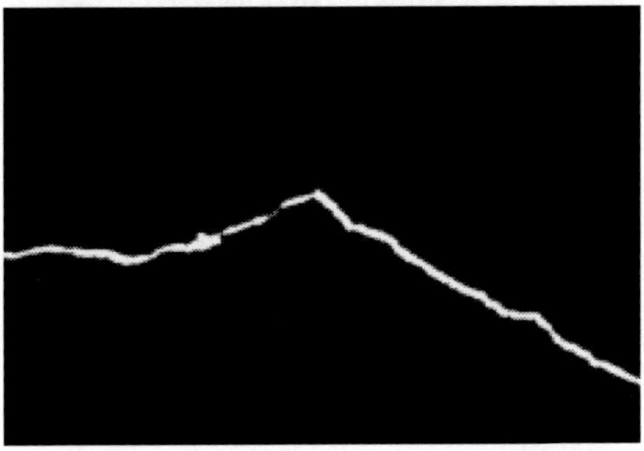

Figure 15. Binarization negated.

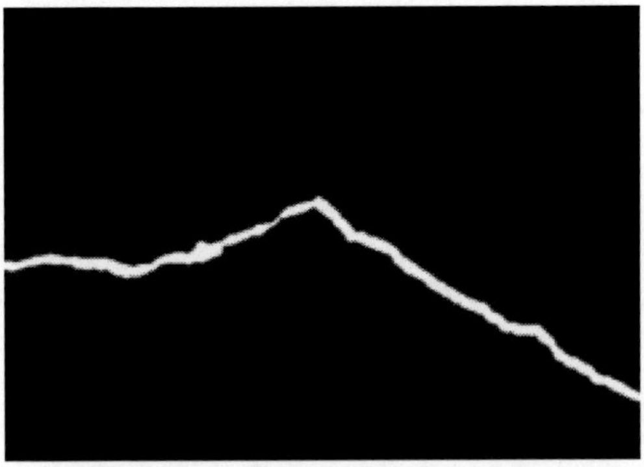

Figure 16. Expansion erosion image.

4.3. Quantitative Analysis Image

After matlab image processing in order to study their difference with the original image, and image is obtained some information such as cracks length, width, etc., we need to quantify the processed image; in this experiment we are mainly to be crack width and area analysis, flow chart 17.

Figure 17. Quantitative Analysis flowchart.

Throughout the quantitative analysis process, you first need to extract the cracks (cracks or original) in Photoshop (referred to as PS) carried matting, and conducting error artificial matting selected when the original image analysis as a standard, other original image contrast after the image processing and analysis. After selecting the constituency to constituency converted into the working path, and export to Adobe Illustrator (abbreviated AI), as shown in Figure 18.

Figure 18. Importing AI cracks roadmap.

Finally, we will export the AI path for the DWG format, in the Auto CAD in the quantitative analysis of the crack. Since the original and extracted crack FIG sample sizes, and in the PS and AI in image analysis are in pixels, in theory, the first decomposition extract the image size should be 1/4 of the original image, but taking into account during operation cause some image information is lost in this selected based on a scale to determine the proportion of original and processed after certain length feature points of the image. In Figure 19, for example, can break point after two pictures and identify the figure as a proportion of the ratio of the crack distance after picture and image processing. Practice has proved that after processing the image with the original image ratio is generally less than 0.25, which is generally close to 0.2412.

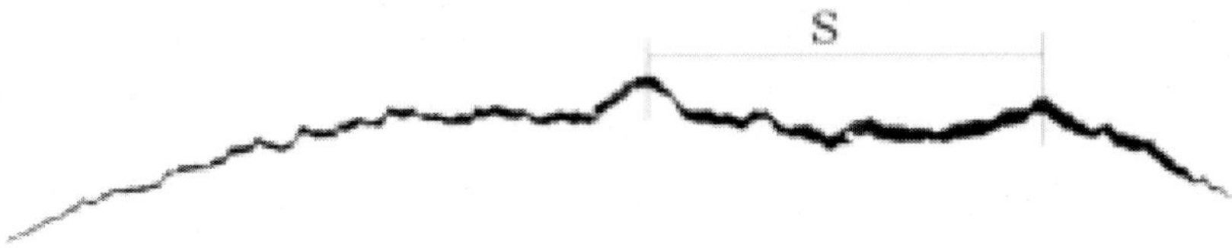

Figure 19. schematic feature points.

In addition, we also need to find a scale and the actual image pixel size; the proportion of pixel values is also some special points by actual distance corresponding to the distance determined.

4.4. Crack Test Analysis

In this section, the author of crack for a number of different forms of experiment process and the results are shown below in order to facilitate analysis of the experimental results below 1 unit length = 1 pixel, and not converted into units of length used in peacetime.

Differences in Treatment of Experiment 1

In this experiment, to investigate the effect of different treatments on when the cracks, this experiment Figure 7 crack test, first take haar wavelet decomposition function of crack recognition four times, and then the different post-processing, processing Figure 8-13 shows the effect of the results of each step of treatment was recorded in Table 1.

It should be noted that the last step is to establish the basis of the previous step. In this experiment, mainly through the line binarization, binarization negated, the expansion has corrosion treatment, the post-processing of the source as follows

```
BW1 = imread('X.jpg');

H = im2bw(BW1,0.5);

I = 256-H;
```

subplot(131),imshow(BW1);title('Original image');

SE1 = strel ('diamond,' 7);

J = imdilate (I, SE1);

subplot (132), imshow (J); title ('expanded image');

SE2 = strel ('disk,' 4,4);

K = imread (J, SE2);

subplot (133), imshow (K); title ('after the first expansion of corrosion image');

Table 1. Treated differently analysis table

No	Treatment	Area	Perimeter	D'Max
0	standard	3228.90	1341.78	14.40
1	extract	3486.91	1318.21	15.14
	error(%)	7.99	1.76	5.13
2	Binarization	3586.56	1329.81	15.17
	error(%)	11.08	0.89	5.35
3	Binary negated	3586.55	1329.82	15.17
	error(%)	11.08	0.89	5.35
4	Expansion of corrosion	4541.68	1321.90	16.39
	error(%)	--	1.48	13.8

NOTE: D'Max represents the maximum width of crack.

In fracture identification process, crack wavelet analysis has identified a very high accuracy, in this experiment directly with the crack wavelet recognition of its accuracy in the area, perimeter, and width of aspect can achieve accuracy were 92.01%, 98.28%, 94.87%. Some post-processing mode, such as binary image processing, the image is negated operation, enabling an appropriate increase in the accuracy of the image, as shown in Table 1 in terms of the accuracy of the perimeter to improve a lot, and image post-processing to facilitate our further the study. However, the image processing is not so easy to form, in the cracks of the test process, I tried a variety of ways, it is similar to solve some discontinuity, the noise of the type of problems, but it will cause some in the process new problems, so accuracy is low.

Experiment 2 Select the Effects of Different Wavelet

The main purpose of the experiment 2 is to study the effect of selection Cracking identify different wavelet function, in this experiment, choose a different image 20 crack extracted wavelet function, decomposition level are four. After the result of the establishment in terms of the line of work after binarization path will bring a lot of convenience, can reduce human error, therefore, in this experiment, I use directly after extraction and binarization as a direct result of the recognition result, the expansion of corrosion results for the post-processing result image 21, results shown in Table 2.

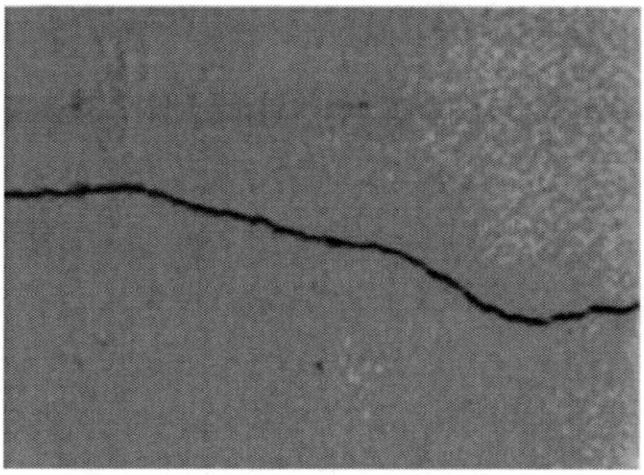

Figure 20. Test Crack 2 picture.

Figure 21. Wavelet recognition crack.

Cracks in the test identification process, selecting different wavelet function to identify cracks in which poor Sym2 wavelet recognition effect, lead to considerable error. Despite the effect of the other wavelet difference, but the difference is not very obvious, after identifying cracks in the area of respect despite larger error, but in girth and width recognition can achieve high accuracy, the width can be controlled in terms of a fundamental error within the pixel length.

Table 2. Wavelet analysis

No.		Direct identification			Post-processing		
	Wavelet function	Area	Perimeter	Dmax	Area	Perimeter	Dmax
0	standard	1962.82	1219.39	7.24	--	--	--
1	Sym2	2284.13	1246.75	6.87	1547.47	1015.52	6.27
	Error (%)	16.37	2.24	5.11	21.16	16.72	13.39
2	haar	2291.59	1395.39	7.23	1872.06	1187.69	6.75
	Error (%)	16.75	14.43	0.14	4.62	2.60	6.77
3	demy	2308.14	1264.65	7.41	1848.67	1195.08	6.76
	Error (%)	17.59	3.71	2.34	5.82	1.99	6.62
4	db-1	2301.01	1398.32	7.23	1880.08	1185.00	6.75
	Error (%)	17.23	14.67	0.14	4.22	2.82	6.77
5	Coif2	2302.22	1257.49	7.43	1850.81	1190.47	6.75
	Error (%)	17.29	3.12	2.62	5.71	2.37	6.77

NOTE D'Max represents the maximum width of crack.

Experiment 3: Different levels of the Same Wavelet Decomposition Function

The main purpose of the experiment 3 is to study choose the same wavelet function, the effects of different levels of decomposition were to identify the effect of the crack, in this experiment, the image 22 using haar wavelet function were 2,4,6,8-level decomposition, the choice after binarization result is a direct result of recognition, choose the expansion of corrosion after the results for the post-processing result image 23, results shown in Table 3.

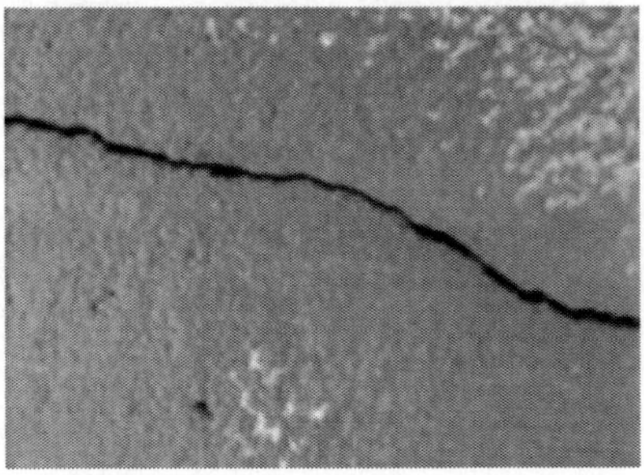

Figure 22. Test crack 3 picture.

Figure 23. wavelet recognition.

Table 3. Different levels of the same wavelet decomposition

No	Break down level	Direct identification			Post-processing		
		Area	Perimeter	D max	Area	Perimeter	D max
0	standard	2475.97	1279.40	9.88	-	-	-
1	Level2	3151.38	1438.71	11.58	2690.77	1225.49	10.37
	Error (%)	27.28	12.45	17.21	8.68	4.21	4.96
2	Level4	3151.26	1438.70	11.60	2690.77	1225.43	10.37
	Error (%)	27.27	12.45	17.41	8.68	4.22	4.96
3	Level6	3151.38	1438.71	11.58	2690.77	1225.49	10.37
	Error (%)	27.28	12.45	17.21	8.68	4.21	4.96
4	Level8	3151.38	1438.71	11.60	2690.77	1225.49	10.37
	Error (%)	27.28	12.45	17.41	8.67	4.21	4.96

NOTE D'Max represents the maximum width of crack.

In wavelet decomposition process, due to cracks experiments were first extracted partial decomposition stage, decomposition level and thus had little impact on the accuracy of the identification. It is worth noting that the test crack 3 with other biggest difference lies in its image acquisition is collected after the rain, due to direct rainwater makes wavelet analysis identified other cracks crack relative error is large, but after it by post-processing still able to get good accuracy, especially in terms of precision area is much higher than other tests cracks.

4.5. Actual Cases

Cracks on image analysis section only is extracted as a reference manual, and its length in pixels for the unit. In this section, we selected two cracks in facades wall art library (Figures 24, 25) for the actual measurement, select multiple control points (26, 27) measuring the crack width and crack width and image processing to give scale measurements were compared. Since the fracture morphology is relatively simple, it did not crack processing of morphology, only to later identify binarized wavelet cracks as the measurement value, the comparison results are shown in Table 4 and 5.

In addition, in determining the length of the actual pixel scale crack, the author first measure certain characteristics of the crack length between the points, and then determine the scale according to the length of the expression after importing CAD. Cracks in test 4, select the upper control points 1 and 4 the upper end of the control point from the actual measuring length 165mm, CAD conversion to a length of 430.53 pixels in length, it is possible to determine the dimensions of 2.61 pixel length = 1mm. 5 cracks in the test, select the cracks start to finish the whole distance, measured 154mm, distance between two points to CAD 433 pixels in length, so the scale of 2.81 pixel length = 1mm.

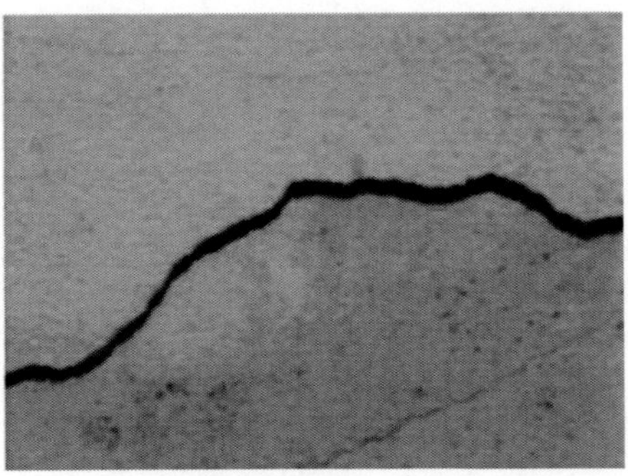

Figure 24. Test cracks 4.

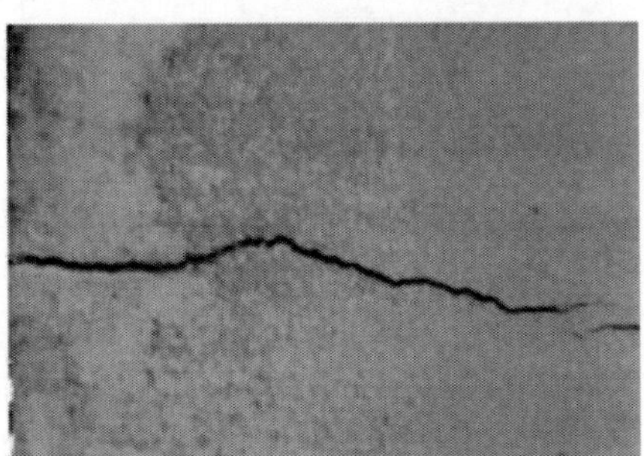

Figure 25. Test cracks 5.

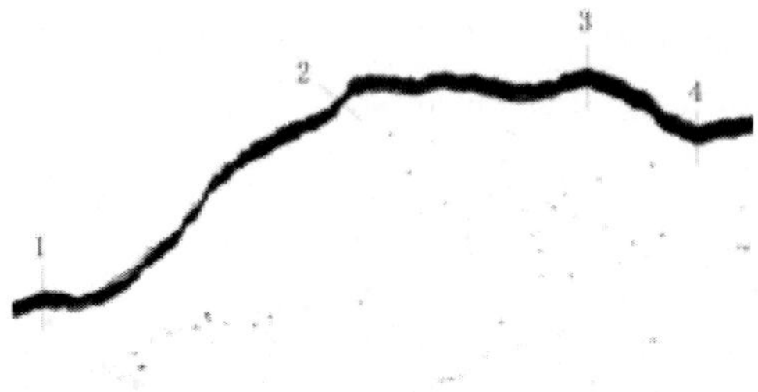

Figure 26. Test crack 4 selected control points.

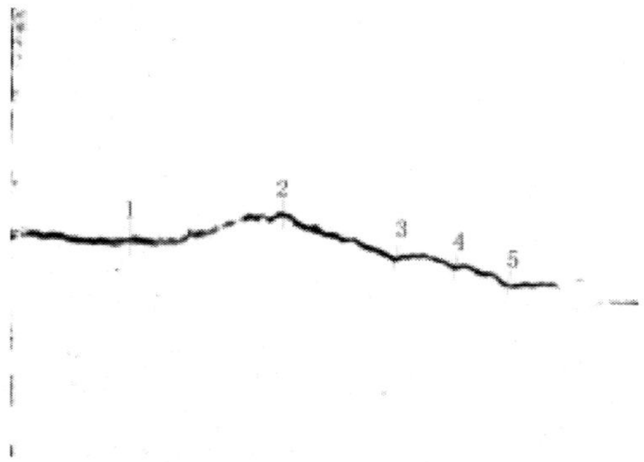

Figure 27. Test crack 5 control points selected.

Table 4. Test crack 4 results

Control Point	1	2	3	4
Actual (mm)	4.2	2.5	5.5	5.7
Measurement (pixels long)	10.44	5.65	14.47	15.07
Measurement Converter (mm)	4.0	2.16	5.54	5.77
Error (%)	4.76	13.6	1.09	1.23

NOTE scale pixel length = 1mm 2.61.

Table 5. Test crack 5 results

Control Point	1	2	3	4	5
Actual (mm)	2.0	1.8	1.7	1.5	1.3
Measurement (pixels)	6.11	4.93	4.34	4.22	3.28
Measurement Converter (mm)	2.17	1.75	1.54	1.50	1.17
Error (%)	8.5	2.78	9.41	0	10

NOTE scale pixel length = 1mm 2.81.

Although the error is greater individual control points, the control points are mainly smaller crack width point. In fact, not difficult to understand, at the same pixel error situation, the greater the crack width smaller error. For most of the control points to achieve good accuracy, and its accuracy can be maintained at substantially more than 90%.

In the actual project we need to find a reference to crack image to determine the relationship between the pixel and the actual size, and this reference is not difficult to choose. To measure the beam crack, for example, reference can be selected beam width or beam higher, as long as the crack image contains beam width or beam width and other information, it is easy to build a scale, which measured the crack-related information, significant savings in manpower and resources. In addition, we also found through the practice of using wavelet analysis is simple to operate, for a lot of cracks requires only a simple recognition can achieve high accuracy, this can reduce the late programming. Finally, in the wavelet choice difference different results due to the selection of wavelet analysis was not particularly evident in the actual operation generally do not need to consider the use of that kind of wavelet to analyze, or you can use two or three different kinds of wavelet analysis, the Results difference within a certain range can be considered as the recognition result is valid. Even if not very professional engineering and technical personnel knowledge, it is possible to operate, which is undoubtedly conducive to practical engineering application practice. Thus, the wavelet analysis has great potential for development and application prospects.

CONCLUSION AND OUTLOOK

Has great potential for development of machine vision-based health monitoring method in inconvenience only need to reach the place by remote camera can analyze the structure of some status, saving manpower and resources. Wavelet Analysis in fracture identification has simple, small steps, calculate fast and high precision characteristics, wavelet analysis in this paper, through the cracks of different wavelet analysis, the following conclusions

1. The use of crack wavelet analysis identified with high accuracy, in terms of the crack width measurement accuracy can reach 1 to 2 pixels, able to meet the needs of practical engineering application.
2. For the effect of crack image processing depends first on the clarity of the image itself. Outline a clear, better image contrast relatively easy

to identify it, and the error is smaller. Pixel image containing a large amount of information, greater accuracy, but its operation is slower.

3. In recognition function wavelet choice to select a different function in spite of some differences, but the difference in fracture identification process is not great.
4. Although the fracture identification process can solve some problems like discontinuity, the use of appropriate post-processing approach can improve the accuracy of image recognition, but it can also produce some new errors. Under the conditions meet the accuracy, identify patterns as simple as possible, the number of image processing, the better.

Although wavelet identified cracks can have a good accuracy, and can identify a variety of cracks, but there are many problems, where we can make some follow-up research prospects as follows

1. The wavelet analysis research has a single, continuous cracks have good recognition results, but more for the number of cracks, fractures form a complex image recognition results are not satisfactory, you can follow this regard.
2. Although the crack width and girth error is very small, but in terms of the error area is still quite large, and post-processing of large cracks extract difficulty in this regard is a research direction.
3. This is due to technical problems and time, in the course of quantitative analysis is to study the physical characteristics of the crack plane that crack width, perimeter, area, and did not go to study spatial characteristics cracks, cracks and other information such as depth, Crack spatial characteristics is also the direction of future research.

ACKNOWLEDGMENTS

The authors deeply acknowledge the financial support from HuaQiao University, Fujian, P. R. China under the HuaQiao Scientific Research Foundation for Talents plan.

REFERENCES

[1] Zhu Hongping, Yu Jing, Zhang Junbing, structural damage and power Detection Monitoring Health Status and Prospects, *Engineering Mechanics,* 2011-2,28 (2), pp. 1-11.

[2] Zhu Hongping, Yu Jing, Zhang Junbing, structural damage and power Detection Monitoring Health Status and Prospects, Engineering Mechanics, 2011-2,28 (2), pp. 1-11.

[3] Bear seashells, Zhang Junjie. *Super high-rise structural health monitoring system overview structural engineer*, 2010-2, vol 26 (1). 144-150.

[4] Sun House of the status quo and development, Zhu Tong, Yang Guanghui, Bian Mei, Fengjun Fu, health monitoring of civil engineering, *Shandong Institute of Architectural Engineering*, 2006-2,21 (1), pp. 81-85.

[5] Charles R, Farrar, Keith Worden, An introduction to structural health monitoring, *Philosophical Transactions of The Royal Society,* December 2006, pp. 303-315.

[6] Study situation Li M, east Wei Yanhua Iraqi Civil Engineering and health monitoring systems. *Advances in Mechanics progress*, 2008-3, 38 (2). 151-166, in Chinese.

[7] JMWBrown john, Structural health monitoring of civil infrastructure, Philosophical Transactions of The Royal Society, December 2006, pp. 589-622.

[8] Li M, Li Dongsheng, Civil Engineering structural safety assessment, health monitoring and diagnosis are reviewed. Earthquake Engineering and Engineering Vibration, 2002-6, 22 (3). 83-90.

[9] Kammer DCSensor Placement for On orbit Modal Identification and Correlation of Large Space Structures, Journal of guidance, Control and Dynamics, 1991, 14 (9), pp. 251-259.

[10] Koh CG, Zhao SL, Liaw CY, Non-destructive parameter identification of structures, *SPRINT Society Optical Engineering*, 2002, pp. 20 ~ 27.

[11] Zhang Qinghua, Bmv Delete A, Wavelet Network s, *IEEE Trans On Nellml Networks,* 1992,3 (6), pp. 889-898.

[12] Michacl A Sutton, Stephen Rmcneill, Jeffrey D Helm, et al. Advances in Two-Dimensional and Three-Dimensional Computer Vision, Topics in Applied Physics, Springer Verlag, 2000, pp. 323 ~ 372, in Chinese.

[13] Shen Yu, binocular stereo vision based structural deformation monitoring technology, Harbin Institute of Technology master's degree thesis project, 2011, 6.

[14] Qin rights, health monitoring of bridge structures, *Chinese Journal of Highway,* 2000,13 (2), pp.37-42.

[15] Yang Jie, current research and development at home and abroad, such as Wu, dam safety monitoring, Xi'an University of Technology, 2002, 18 (1), pp. 26-30.

[16] GU Chong, Wu such as, structure and knowledge engineering dam safety monitoring expert system technology Supervision in Water Resources, 1998 (1), pp. 36-40, in Chinese.

[17] Health Status Monitoring Weibao Li, Su Xiaohui, bridges analysis, Northern Jiaotong, 2008 (1), pp. 122-125.

[18] Deng Julong, Control problems of grey systems, *Systems and Control Letters,* 1982,5 (1), pp. 288-294.

[19] Bear seashells, Li Zhiqiang, structural health monitoring, *structural engineers,* 2006-10,22 (5), pp. 86-90.

[20] Juan, Yang Na, Yang Shan, applicable to health monitoring system heritage building structures, Beijing Jiaotong University, February 2010, pp.100-104.

[21] Wang Kaifeng, Zhang Xiedong, civil engineering structure damage identification research, roads, 2008-1, the first phase, pp. 154-159.

[22] Lin Tan, South China University of Technology PhD thesis, based on structural damage identification method for dynamic reliability of fingerprint research, 2010-5-4.

[23] The objective function and algorithm Yang Zhichun, Wang Yue, Li Bin, Lau Kong-wah, structural dynamics finite element model, *Journal of Applied Mechanics*, 2009-6,26 (2), pp. 288-296.

[24] Limin, Sun Chi, Tan Dan Hui, Zhang Qiwei, to just study and application of the status quo. China's large-span bridge structural health monitoring system. 17th National Bridge Conference Proceedings (the book), 2006. 664-670.

[25] Li Wanrui, structural damage identification method and timing analysis model updating based on research, Lanzhou University of Technology PhD thesis, 2013-4.

[26] Zhang Yu-chi, structural damage identification theory and neural network based on digital convergence, Southwest Jiaotong University master's thesis, 2007-1.

[27] Staff, Wei Fu spin. Model updating and damage diagnosis. Science Press, 1999.

[28] Artificial Wangxiu Xin, Li Aiqun, neural network used in structural health monitoring project discussion, *the special structure*, 2007-6, vol 24 (2), pp. 96-99.

[29] Following Ge Branch, Review of Qiu Yuhui, bility, Pu Guolin, genetic algorithms, computing and applied research, 2008-10, vol25 (10): pp. 2911-2916.

[30] Edge Xia, Miliang Theoretical Research on Genetic Algorithm and its Application, Computer Application Research, 2010-7, vol 27 (7): pp. 2425-2434.

[31] Li Ge, Qin rights, Dong Cong, select Suspension sensor monitoring system using genetic algorithms in optimal distribution, engineering mechanics, 2000-2, vol17 (2): pp. 25-31.

[32] Sunzeng Shou, Han Jiangang, Ren Wei new research progress monitoring structural damage based on wavelet analysis, *Earthquake Engineering and Engineering Vibration*, 2006-4, vol 25 (2): pp. 93-98.

[33] Li Jianping, Chen Tinghuai, Xu asked, the Chang Wan Ping, to see the development of wavelet theory from literature analysis, Chongqing University, 1997-1, vol 20 (1): pp. 82-87.

[34] Maas, H.-G., and Hampel, U. (2006), Photogrammetric techniques in civil engineering material testing and structure monitoring, *Photogram. Eng. Remote Sens*, 2006, 72 (1), pp. 39-45.

[35] RC Gonzalez, RE Woods, Digital Image Processing (Second Edition) [M], Ruanqiu Qi, Yu-Chi Nguyen, translated. Beijing Electronic Industry Press, 2003.

[36] Zhai Lei, Dongshou Ping, Ma Honglian, digital image processing technology development status and prospects, the 19th National Computer Technology and Applications (CACIS) Conference Proceedings (the book), 2008, pp. 387-392).

[37] Yoji Ling Wang Kai. ViusalC ++ digital image acquisition, processing and practical application Beijing: People's Posts and Telecommunications Press, 200.3.

[38] Zhu Chunyan, Rong Lihong, development and application of Comrade Min, Meng Yaohua, computer vision technology, *Heilongjiang Science and Technology Information,* 2013, vol 29, pp. 174, in Chinese.

[39] Zhangguo Xuan, the automotive industry using machine vision systems, laser and Optoelectronics Progress, 1987, vol 2, pp. 43.

[40] Section peak PROCEEDINGS Lei Xiaofeng, Wu Li Zhao, Tan, machine vision technology and its application review, *Automation Expo*, 2002, vol 3, pp. 59-61.
[41] Xiong Lirong; Dingyou Chun; LIU Jianying; machine vision technology in the agricultural field, *Hubei Agricultural Mechanization*, 2004, vol 4, pp. 23.
[42] Orthopaedics, Shen Hongyan, machine vision technologies in industrial inspection review of the use of electronic technology and software engineering, 2013, vol 22, pp. 111.
[43] Raschke S A. Computer Vision for Experimental Soil Micro- mechanics and Soil Characterization Michigan: The University of Michigan, 1996.
[44] YF Ji, CC Chang, Nontarget Stereo Vision Technique for Spatiotemporal Response Measurement of Line-Like Structures, *journal of engineering mechanics asce/june 2008/*466-474.
[45] Ye expensive, such as, Zhou Qingsong, Lin Xiaowei, based on the surface crack width measurement of digital images, Highway and Transportation Research, 2010-2, vol 27, 2, pp. 75-84.
[46] Guards, Caolong Fei, who Yanxia, Dongrong Zhen, Chen Shaolei cracks feature extraction of concrete structures, Wuhan University of Technology, 2011-11, vol 33-11, pp. 72-75.
[47] Nie Honglin, Hu Wu-sheng, the walls crack deformation monitoring studies based on digital image processing technology, the mapping project, 2012-8, vol 21 (4), pp. 61-69.
[48] Identification of materials from construction site images using content based image retrieval techniques, *Computing in Civil Engineering,* 2005, 1-12.
[49] ZhiQiang Chen, Tara C. Hutchinson, Urban Damage Estimation Using Statistical Processing of Satellite Images.

In: Robotics: New Research
Editor: Daniel G. Fisher
ISBN: 978-1-63485-967-7

Chapter 6

UPPER ARM EXOSKELETONS – WHAT SPECIFICATIONS WILL MEET USERS' ACCEPTABILITY?

Soumya K. Manna and Venketesh N. Dubey*
Faculty of Science and Technology, Bournemouth University, Poole, UK

ABSTRACT

There have been many upper arm exoskeletons developed in the last two decades however none of these have met the expectations of the real users. Most of these designs did not go beyond the laboratory environment and were mainly used either as a research tool or as experimental devices for a very specific application. The majority of these devices were platform-based that was one of the main inhibiting factors why users found such devices too cumbersome to use. Even some of the portable arm exoskeletons did not find much favourable acceptance amongst the user groups because they were too heavy or difficult to put on and operate. This opens up a question, what specifications of upper arm exoskeletons will meet users' acceptability? This article presents a systematic review of upper arm exoskeletons to find out under what clinical conditions use of such devices may be beneficial, what could be the technical requirements and what user interface must be provided to enhance their acceptability. This paper looks into all major aspects of design related issues of upper arm exoskeleton that existed and the challenges ahead to overcome from technical as well as clinical application point of view. Whether arm exoskeletons can be used for stroke patients, Parkinson's or patients with weak musculature in assistive mode or for rehabilitation training? Also for those patients who suffer from different neuromuscular disorders have stiff muscles resulting in arm spasticity that doesn't allow free movement of the joint therefore the exoskeleton design should be adaptable to such clinical needs as well. The physiological parameter should be set as a standard for measuring patient status as the rehabilitation progresses and for bio-feedback. On the technological aspects the design should considers arm's degrees of freedom, dependency of joints and control regime to justify the optimal design performance. It should be a portable system with active and passive joints with gravity compensation provision to have low energy consumption. The type of actuator selection is one of the important criteria for the acceptability of such devices. It must also consider reconfiguration of the exoskeleton for misalignment, singularity, redundancy in design and weight. There are a large number of sophisticated control algorithms existing but most of them are not adaptive to safety measures and for the ease of operation. In addition to the above, the design must also consider vitally important user interface particularly with the advent of inexpensive electronics gadgets such as Mayo Armband, Google glasses, Eye Tracking systems and Kinect for motivating user rehabilitation.

Keywords: arm exoskeleton, portable, rehabilitation device, hybrid mechanism

INTRODUCTION

Exoskeletons or wearable skeletons are people-oriented robots designed to be worn for training and assistance. These robots are designed based on the function and shape of the human body and the user is able to control the robotic limbs. These exoskeletons can assist in walking, running, jumping or even lifting objects one would normally not be able to lift. These examples are only some of the most basic ways that engineers hope one day these robots will be used for. It may be attached to different parts of body such as upper limb, lower limb, and wrist or as the full body vest. However, in this paper we are only looking at upper arm since arm mainly contribute to the activities of daily living. For the last three decades there have been a number of research projects on human arm exoskeleton, it started with the concept of supporting military personal in warzone for undertaking strenuous works. However, as the time moved on, the area of application has been expanded to a large number of diverse applications. Broadly it can be divided into three major areas firstly; it can be used as an assistive device for paralysed and elderly people secondly, as a therapeutic device for patients suffering from different neuromuscular weakness and finally, robotic exoskeletons that could be used as a powerful muscle support for military personal for lifting under hazardous loading conditions. So the design and development of such systems are driven by different requirements depending upon the application.

The main area of research on exoskeletons has been motivated and directed towards medical application as therapeutic devices [1, 2]. It is not fallacious to say that exoskeletons have been mostly used as dedicated rehabilitation devices after stroke. Stroke is a life-threating event that affects not only the person who is victim, but their family members and the caregivers. It is seen by more than half of the stroke survivors as being worse than death. Paralysis is a condition caused by complete loss of muscle function of one or more muscle groups. In such cases, patients don't have any control on physical movement of the affected part. Also different cognitive problems may occur due to stroke and neuromuscular deficits like problems with concentration, memory and visual perception [3]. As a result, the sufferers are not able to do activity of daily living known as apraxia. In some cases, patients are not able to do the work on one side of their body known as hemiplegia because of the effect of spatial neglect. In some cases, the patients may be suffering from spinal cord injury or nerve diseases and may also suffer from muscle power loss. On the other hand, human muscles may get stiffer and rigid which may restrict patient's free movements of their limb such as arm spasticity and uncontrolled arm movement in Parkinson's disease. In general, people will lose their mobility, way of communication as well as their strength; therefore, they will require some external support to do their activities [4]. It has been proven that if the patients are under rehabilitation for several months after stroke, their active range of motion as well as muscle strength can increase significantly [5]. Sooner the rehabilitation training started the more likely they are to regain the lost abilities and skills. It is common to start rehabilitation as soon as 24 to 48 hours after stroke during the acute phase of hospital stay. A repetitive and early training session will improve the neuro-motor functions. The training includes all orthopedic and neurological lessons that would be compatible with the human muscle movement. The treatment for the stoke patients will be better if the training session is continuous and as many times as possible in a day. The rehabilitation training is generally performed by physiotherapist but patient has to confront a lot of socioeconomic cum technological problems [1, 6]:

1. First of all, it has been found that a physiotherapist can perform the training for 8-10 hours a day and can provide service to 2 to 3 patients with a very high cost. According to the Stroke Association of UK, the numbers of stroke affected people are increasing day by day which in turns also increases the demands of intensive therapy.
2. In the care homes, there is a shortage of sufficient number of trained staff. Therefore, normal people have to perform those activities which may not be effective and may cause some adverse effects to the patient. They may even omit certain exercises which are essential to the patients.
3. The time of training is not adequate due to the fatigue of therapists. The patients do not get repetitive and accurate movement sessions in case of a manual intervention. On the other hand, exoskeleton could potentially provide rehabilitation training for a longer period without any break even without the presence of therapist.
4. For remote areas where it is difficult to reach or due to unavailability of such therapist, robotic assisted therapy could be implemented and controlled remotely. Any trained person can offer rehabilitation training to a patient with the help of exoskeleton system from their normal place of work.

Considering the time, cost, repeatability, reproducibility and accuracy, robotically assisted rehabilitation process could be better than human assisted training. The purpose of exoskeleton in human rehabilitation is to strengthen motor skills, mobility training, constraint-induced therapy, and a wide range-of-motion therapy.

Robotically Assisted Therapy

Initially end effector based systems had been used for physical therapy. It only connects the distal part of human arm with a system that looks like a serial manipulator. The movements of end effector can change the angular positon of the arm, since the whole arm is connected to the system all other segments will automatically move with it. It is less complicated and easy to design since it does not take care of the structure of arm. But the main problem with this, it cannot initiate a particular movement to a specific joint. Depending on the configuration sometimes it may create risk to patient's arm. Notable systems in this category are MIT MANUS [7], CRAMER [8] (Figure

1). Since our consideration is whole arm exoskeleton such systems are not our focus in this paper.

MIT MANUS [7] CRAMER [8]

Figure 1. End effector based robotic rehabilitators.

It is however not an easy task to design a human arm exoskeleton; because of its complexity, several degrees of freedom, and dependency of joints and the associated control architecture. There is also a lack of technology advancement in design related issues such as shoulder joint implementation and joint offset compensation. The design of the structure does not follow any standard guidelines. The existing models still have a lot of ergonomic challenges [9]. It is still not established how many degrees of freedom is optimal. Most of exoskeletons developed so far are platform-based or stationary. But the market survey [10] suggests user would like to have a portable system as it will be helpful in assistance as well as training for day to day activities. All exoskeleton models should be verified from mechanical as well as clinical point of view. The most design implementations of exoskeleton have been for stroke sufferers but it should also be applicable to the patients suffering from other neuromuscular diseases. For a portable system, energy consumption is another critical issue, there are only a few exoskeleton devices that are commercially available and those too are mainly therapeutic devices. Also the cost of such kind of devices is too prohibitive for normal people to afford. Adaptation of new technology in terms of active and passive exoskeletons is another consideration; sometimes the designs are so complex that patient acceptance rate is quite low [11]. We therefore aim to propose an exoskeleton device where most drawbacks could be compensated. The type of actuator selection is one of the main criteria which decide how heavy the exoskeleton would be. A large number of sophisticated control algorithms have been implemented but these are not adaptive in nature. Safety features should be built into the system because the system is to be used by human beings. In addition to the above criteria, other vital parameters need to be considered is to design and configure an exoskeleton such as alignment, redundancy, singularity, actuation and portability of the exoskeletons. This paper aims to address some of these disparate design requirements and develop a comparative study of different existing exoskeleton models.

TYPES OF EXOSKELETON DEVICE

This particular area of research needs a complete understanding of human movements and their mechanism, involving the anthropometric data, its kinematics and dynamics. We are confining our research to the medical application of exoskeleton. Mostly two types of design have been developed as shown in Figure 2. The first one is a ground-based exoskeleton system which is attached to a base platform [12, 13]. The second one is a body-based portable/wearable device [14, 15]. If the exoskeleton has to be used as an assistive device, this has to be a body-based system. Increasing the degrees of freedom of the structure would make it more flexible and redundant like human arm but at the same time it will also increase the complexity of the system so an optimal design has to be struck.

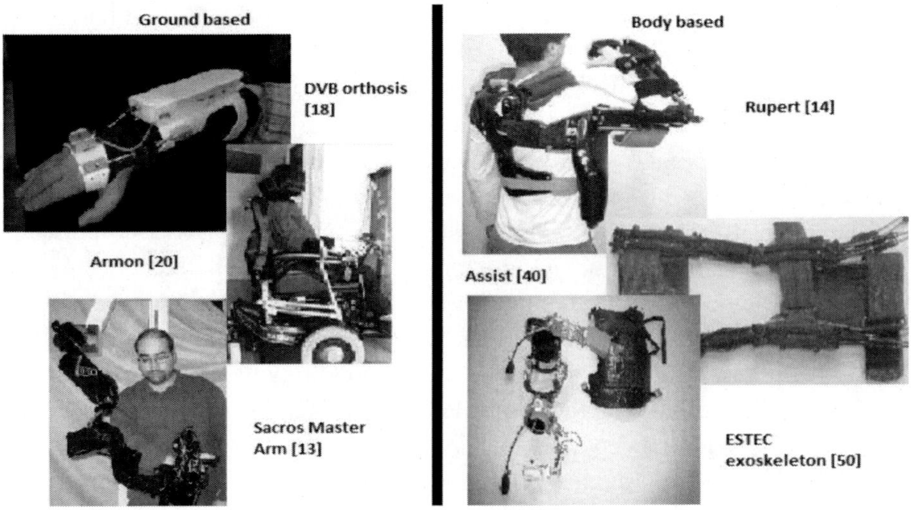

Figure 2. Basic exoskeleton system (ground and body based).

Upper arm exoskeletons are generally used for two main purposes for medical applications; (i) for improvement of muscle strength through physical therapy and (ii) to support in activity of daily living (ADL). However, a survey conducted suggests that most of the developed exoskeletons are dedicated to physical therapy [1] and the number of exoskeletons used for ADL is quite low. Assistive exoskeleton devices could be used for the paralysed people or those who have lost their arm movement completely. These devices should be medically suitable to be worn by the patients, user-friendly, cost effective, multi-functional and safe. The main consideration in designing such devices is portability since the energy source has to be carried by the user all the time, therefore, the mechanism requirement should be such that it will consume less energy. If the device is meant to have a flexible structure it will have several degrees of freedom, this entails a lot of actuators which will increase the weight and size of the exoskeleton. Devices like HAL [16], WOTAS [17] fall under this category (Figure 3). Sometimes such devices have been installed on wheelchair; MUNDUS [19], MULOS [21], shown in Figure 3, for patients who do not have the ability to walk, in such cases the patients do not have to carry the device so can carry out basic ADL without worrying about its size and weight of the exoskeleton.

The other type of exoskeletons is called therapeutic device which are generally platform based, the question is whether such devices should be a complex system installed in hospitals or care centres, or it may be used in a home setting with simple and easy to use configurations. Mostly therapeutic exoskeletons are installed in hospitals because of its complexity and cost bearing factors but the demand is increasing day by day for personal installation. The general preference from medical professionals is not to have too complex system but should be user friendly and less expensive. So far upper arm exoskeletons have been used for the rehabilitation of acute and mild stroke sufferer but there are also other neuromuscular conditions like spinal cord injuries and Parkinson's diseases where use of exoskeleton has been reported [17]. Depending on therapeutic applications exoskeletons can be further divided into three categories such as active device, passive device and haptic device as shown in Figure 4.

Active exoskeletons are basically considered for active muscle cum joint movement for chronic stoke patients who have already lost their muscle strength completely. Here the patient participation is less compared to exoskeleton system because patients neither have the strength nor the control on their movement. Therefore, whatever movements patients want to take are already predefined and loaded in the system prescribed by rehabilitation expert, system will move at different frequency level according to the prescribed motion. Most of the existing devices in the market are able to provide active movement but the exercises are called passive because the patient does not participate actively. However, when patients initiate exercises it is called active exercises thus there is a conflict in meaning between active device and active exercises -active devices generally induce passive exercises. The weight as well as complexity is higher in the active devices because of large number of actuators used to control, as a result the cost of such devices is comparatively higher. Since they use several actuators for executing different movements the system consumes more energy and requires a reliable and higher energy source, it conflicts with the portable exoskeleton requirements. Some active exoskeletons in this category are Armin [11, 22], Exorn [15, 23].

Passive exoskeletons are generally used by the patients who still have some strength left in their arm but not enough to keep it at a particular position. In this case a patient can react or may give some resistive force to oppose the motion. A supporting device in the form of an exoskeleton may be there to provide some assistive force to help the patients during the movement or may resist if the arm motion is going beyond a certain trajectory. The system may use brake or spring actuated mechanism to apply resistive force. Such exoskeletons do not use actuators or have minimum number of actuators but the exercise is called active because patient or user initiates the movement and its being supported by exoskeleton. Passive exoskeleton systems are less energy consuming device therefore could be used as a portable device, it's less expensive because of less number of actuators. Device like MEM-MRB [24] is an example under this category, Figure 5.

EXOSTATION [25] can be regarded as external frames that can provide information on different muscles and joints but are not able to provide either active movement or resistive force. It is another category of exoskeleton where the user will interact with some kind of position and force sensors to be used as a feedback system for collection of sensory data. It has been also used as a master to control an exoskeleton system remotely by measuring the arm movement of human [25]. It does not consume significant amount of energy for operating sensors, Figure 5.

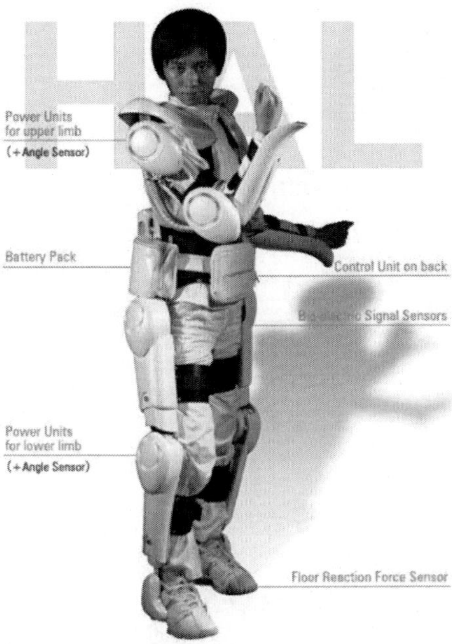

HAL [16]

WOTAS [17]

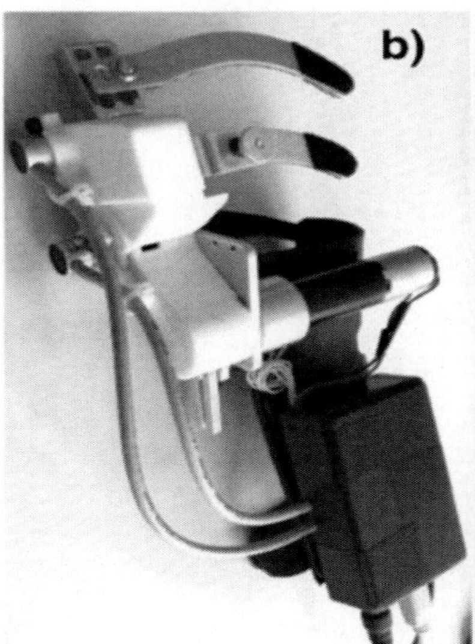

MUNDUS [19]

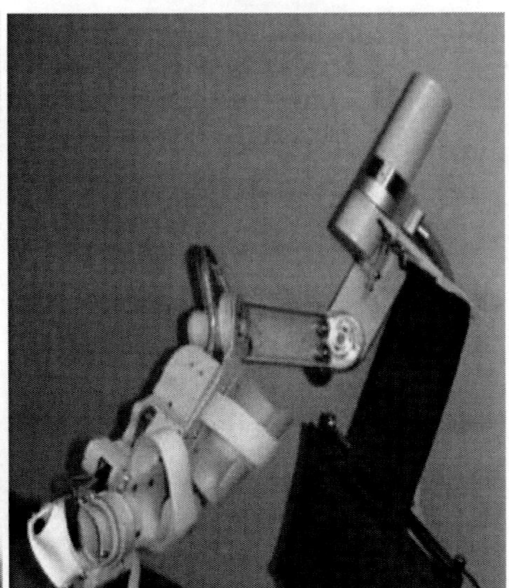

MULOS [21]

Figure 3. Exoskeleton for ADL.

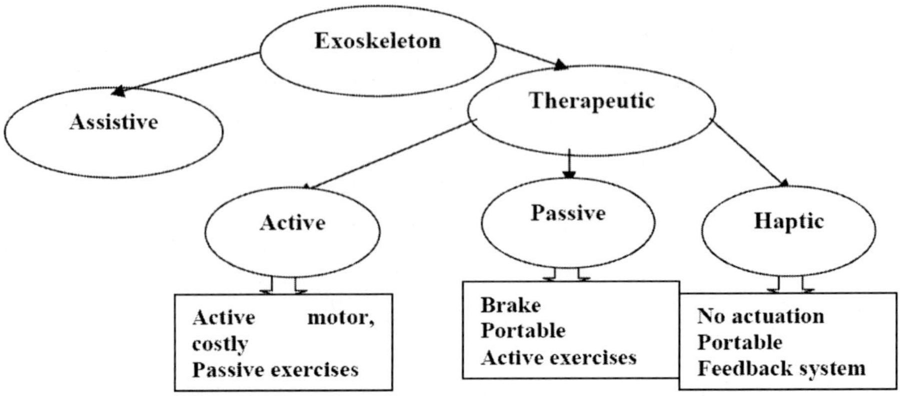

Figure 4. Exoskeleton classification for assistance and training.

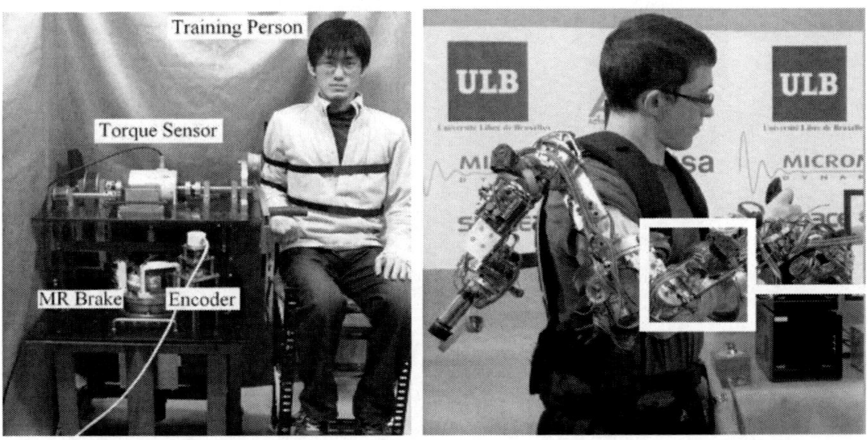

MEM-MRB [24] (Passive) EXOSTATION [25] (Haptic)

Figure 5. Exoskeletons for physical therapy.

In order to consider the capabilities of human arm that could be implemented in exoskeleton design the arm motion can be broken down in to simpler terms as shown in Figure 6. However, a combination of these motions could result in a complex motion sequence that may be difficult to achieve in a portable arm exoskeleton. It might be worthwhile to look at the comparative design features of existing exoskeleton models as presented in Table 1 together with their control architecture to lay out the design specifications that will meet users' acceptability.

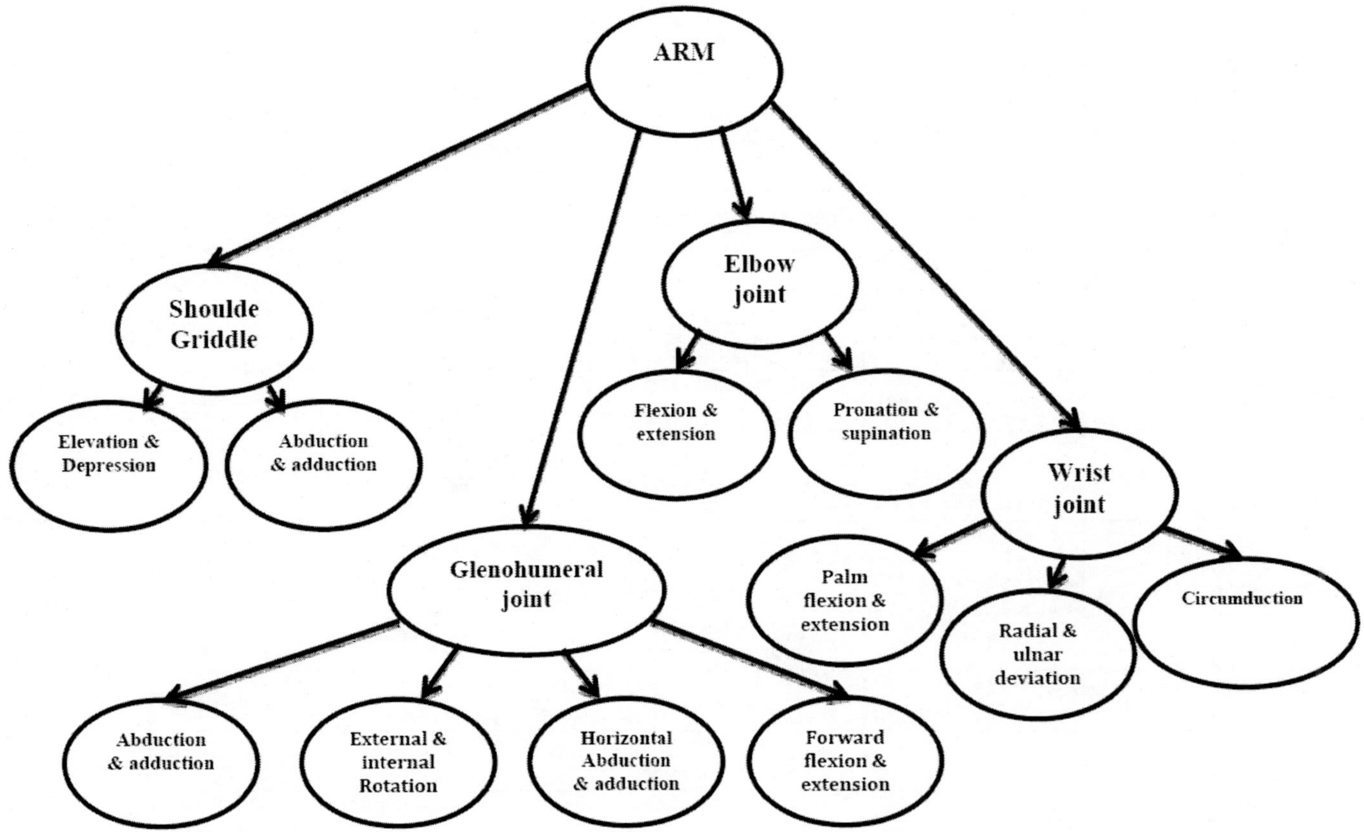

Figure 6. Motions of human arm.

Table 1. Exoskeleton designs with control architecture

Exoskeleton Design	Actuating system	Degree of Freedom	Attached to	Sensors and Feedback	Control architecture	Portability	ADL/ Physical Therapy
Arm-in [11, 22]	Electric motor	6	Shoulder, elbow, forearm wrist	Force and Position sensor	Labyrinth control scheme	No	Physical Therapy
Sarcos Mater arm [13]	Electric motor	7	Shoulder, elbow, forearm wrist	Force sensor	independent PD servo control	No	Physical Therapy
HAL 5 [16]	Electric motor	8	Human body	Bioelectric sensor, angular and acceleration sensors, floor reaction force sensors	Cybernic autonomous Control	Yes	ADL
MGA [26] exoskeleton	Electric motor	7	Shoulder, elbow and forearm	Force feedback and torque	Admittance control and impedance control	No	Physical Therapy
RehaBot [27]	Electric motor	7	Human body	Position and kinematic data controlled	Tele-rehabilitation	No	Physical Therapy
ExoRob [28, 29]	Electric motor	5	Elbow joint and wrist joint	Force feedback	Sliding mode control	No	Physical Therapy
MEDARM [30]	Tendon driven	3	Shoulder, elbow, wrist	Position and kinematic data controlled	Independent joint control	No	Physical Therapy
ShouldeRO [31]	Bowden cables	2	Shoulder joint	Angular position data	Not defined	No	Physical Therapy
NEUROEXOS [32]	Hydraulic drive	3	Shoulder joint	Flow controlled	agonist-antagonist mechanical actuation system	No	Physical Therapy
Multiple Joint Robotic Arms [33]	Ultrasonic motor	4	Shoulder, elbow, wrist	piezo-deflection amplitude	Independent joint control	No	Physical Therapy

Exoskeleton Design	Actuating system	Degree of Freedom	Attached to	Sensors and Feedback	Control architecture	Portability	ADL/ Physical Therapy
Skeleton Arm [34, 35]	Electric motor and tendon driven	6	Human arm	Pressure and Kinematic data	Kinematic and force based control	No	Physical Therapy
BONES [36]	Pneumatic	4	Shoulder and elbow	Joint angles and cylinder pressure	Impedance control	No	Physical therapy
Dampace [37]	Hydraulic actuator	4	Shoulder and elbow	Joint angles and torque	Passive system	No	Physical therapy
Limpact [38]	Rotational hydroelectric actuator	4	Shoulder and elbow	Joint angles and torque	compliant impedance control and stiff admittance control	No	Physical therapy
Armeospring [39]	None	7	Shoulder, elbow, wrist and finger joint	Joint angles and grasp force	Java Therapy*	No	Physical therapy
ASSIST [40]	Rotary-type pneumatic actuators	1	Wrist	Joint angle	EMG and torque based	Yes	ADL
Pneu-Wrex [41]	Pneumatic	5	Shoulder, elbow and finger joint	Joint angles, grasp force	PD force control	No	Physical therapy
Intelliarm [42]	Electric motor	10	Shoulder, elbow, wrist and finger joint	Joint angles and torque	VR based impedance control	No	Physical therapy
MUNDUS [19]	Electromagnetic DC brake	6	Shoulder, elbow, wrist and finger joint	sEMG, Button, eye-movement or brain-computer interface	BCI based control	Yes but wheelchair mounted	ADL

Table 1. (Continued)

Exoskeleton Design	Actuating system	Degree of Freedom	Attached to	Sensors and Feedback	Control architecture	Portability	ADL/ Physical Therapy
HEnRiE [43]	DC brushed motor	6	Shoulder, elbow, wrist and finger joint	End-point torque, position and velocity and joint angles and end-point force	Admittance control	No	Physical therapy
Gentle/G [44]	DC brushed motor	9	Shoulder, elbow, wrist and finger joint	End-point torque, position and velocity and joint angles and end-point force	Admittance control	No	Physical therapy
ARMOR [45]	Electric motor	8	Shoulder, elbow, wrist and finger joint	Joint angles	EMG based control	No	Physical therapy
SUEFUL-7 [46]	DC servo motor	7	Shoulder, elbow, wrist and finger joint	sEMG, Joint forces	EMH based fuzzy control	No	Physical therapy
MIME-RiceWrist [47]	Electric motor	10	Shoulder, elbow, forearm and wrist	Force Sensor	impedance-based force control	No	Physical therapy
UMH [48]	Pneumatic	6	Shoulder, elbow, forearm and wrist	Joint torque	Impedance control	No	Physical therapy
RUPERT [14]	Pneumatic	5	Shoulder, elbow, wrist	Joint torque and actuator pressure	Impedance control	Yes	ADL
Salford Rehabilitation exoskeleton [49]	PAM	7	Shoulder, elbow and wrist	Joint position and torque	PID control	No	Physical therapy
ESTEC exoskeleton [50]	Not selected	9	Shoulder, elbow, wrist	Joint angles	Impedance control	Yes	Physical therapy

Exoskeleton Design	Actuating system	Degree of Freedom	Attached to	Sensors and Feedback	Control architecture	Portability	ADL/ Physical Therapy
CADEN-7 [51]	DC Brushed motor	14	Shoulder, elbow, forearm and wrist	SEMG joint angles and angular velocities and torque	Force base trajectory control and Impedance control	No	Physical therapy
WOTAS [17]	DC motor	3	Elbow, forearm, wrist	Angular velocity, torque	Active tremor suppression control strategy	Yes	Physical therapy
MAHI Exos-II [52]	Frameless DC brushless motor	5	Elbow, forearm, wrist	Joint angles	Proportional control	Yes	ADL/ Therapy
T-WREX [53]	None	5	Shoulder, elbow and finger	Joint angles, grasp force	Java Therapy*	No	Physical therapy
RehabExos [54]	Frameless DC brushless motor	4	Shoulder, elbow and forearm	Joint torques	Force based control	No	Physical therapy
W-EXOS [55]	DC motor	3	Forearm and wrist	sEMG, torque	Fuzzy-neuro control	No	ADL
ARAMIS [56]	DC brushed motor	12	Shoulder, elbow and forearm	Joint angles and torque	Distributed force control	No	Physical therapy
iPAM [57]	Pneumatic	6	Shoulder, elbow and forearm	Joint torques	Admittance control	No	Physical therapy
L-Exos [58]	Electric motor	4	Shoulder, elbow and forearm	Force feedback	Proximal segments motor control	No	Physical therapy
MULOS [21]	Electric motor	5	Shoulder, elbow and forearm	joystick	Angle oriented control	Yes but connected to wheel chair	ADL and Physical therapy
Armon [20]	No actuator	3	Shoulder, elbow and wrist	Joint angles and torque	Zero gravity compensation based system	Yes but connected to wheel chair	ADL

Table 1. (Continued)

Exoskeleton Design	Actuating system	Degree of Freedom	Attached to	Sensors and Feedback	Control architecture	Portability	ADL/ Physical Therapy
DVB orthosis [18]	Magneto-Rheological Fluids	1	wrist	Not defined	DRIFTS control scheme	Yes	ADL
MEM-MRB [24]	Magneto-rheological fluid brake	1	Elbow	Joint angular velocity, torque	PI controller	No	Physical therapy
EXOSTATION [25]	No	7	Shoulder, elbow, wrist	Joint torque	Joint to joint control algorithm	Yes	Haptic device
Hybrid Elbow Orthosis [59]	Hydraulic	1	elbow,	sEMG	EMG based control	Yes	Physical therapy
Exoskeleton Design	Actuating system	Degree of Freedom	Attached to	Sensors and Feedback	Control architecture	Portability	ADL/ Physical Therapy
Hand Mentor [60]	Pneumatic	1	Wrist and finger	Wrist angle, flexion torque	Interdependent joint control	Yes	Physical therapy
Myomo [61]	DC motor	1	Elbow	sEMG	EMG based Haptic and Inertia control	Yes	Physical therapy
ALEx [62]	DC Brushless motor	6	Shoulder, elbow, forearm and wrist	EMG sensor	EE control	Yes	Physical therapy/ ADL
Exorn [15,23]	DC Brushed and brushless motor	10	Shoulder, elbow, forearm and wrist	Joint position	Resolve motion rate control	Yes	Physical therapy/ ADL

*Java therapy: It is a web-based rehabilitation process for repetitive training where patient is able to access website for selecting customized program of therapeutic activities and also receive quantitative feedback of their progress.

EXOSKELETON ATTACHMENT AND JOINT ALIGNMENT

From the above Table it is clear that any model should be based on anthropometric data of human arm. It should also follow the kinematics and dynamics of human joint. In ground based exoskeleton the main advantage is that the motion is very stable and provides maximum torque to the joints but for body-based exoskeleton the length and motor positions are just approximated to the human joints and their alignment cannot be guaranteed. When the motor is actuated it moves the human arm attached to it which make the system dependent on the body configuration. Since actuators are mounted at the joint and it moves with the body it would not get sufficient amount of torque. So to increase the motor torque large motors are required which will increase the weight as well as size of exoskeleton and would not be wearable. One of the major aspects of designing exoskeletons is to align the joints with the human arm when attached. During the joint motion the distal point of exoskeleton will move with respect to that of the human arm in different postures. The attachments will be shifted to a new position as a result the actuator of exoskeleton can move certain part of the human arm which may not be a desirable movement [9]. It may cause injury to the patient such as scratching off the skin. This is a very common problem with body-mounted exoskeletons.

Existing exoskeleton systems have generally impedance or admittance control implemented which are basically providing force feedback to the controller. The force is measured by different sensor during exercises and forwarded to the controller. Also the generated torque is transmitted to a certain joint but because of the misalignment the joint of human arm may not match properly with the exoskeleton joint resulting in an error in force transfer. Also as the exoskeleton system behaves like a mechanical chain connected serially with several linkages an offset will be generated at the end of each distal end. Those consecutive errors will be accumulated and move to the next linkage. Though the offset may be quite minimal for a single joint but the accumulated error could create a major problem in comforts and ergonomic value of the device as it may affect the soft-tissue of arm. Use of spring actuated passive joints could compensate for the variation of length and this can also replace some of the rigid links as elastic element [50]. As an example, in some exoskeleton systems pneumatic artificial muscle has been used as actuators which are compliant with human arm and may also reduce extra burden of passive joint. Malosio et al. [63] have reported that exoskeleton misalignment can offer advantage in an upper arm exoskeleton in terms of movement in rehabilitation and torque decoupling [63]. But from technological point of view the structure should be adaptable in nature and proper alignment should be established for human-robot interactions. The system should have all the safety features incorporated from mechanical, electrical and from software implementation point of view.

ACTUATION SYSTEM

An independent actuator normally can provide one degree of freedom however in human body there are several muscles working together as actuators which contract and expand to give motion to a single joint. Actuation system for arm exoskeleton is different depending on whether the model is active or passive. The type of actuators mostly used for active exoskeletons is electric motor which is easy to control and have high power and bandwidth (refers to its frequency and speed). Though for rehabilitation training bandwidth would not be an issue, weight and actuator size might create constraints. The main problem in using electric motor is high impedance (refers to the restriction of motion during harmonic force) and low torque to weight ratio. In some exoskeletons piezoelectric motors are used in place of DC motors for higher power to weight ratio but it faced with problems of very small displacement and high voltage power sources [64]. A new compensating model has been developed by integrating a parallel elastic element like spring with the actuator, this is called Series Elastic Actuator which not only decreases impendence but also provides stabilizing force in gravity compensation [56]. The options for different types of motor for exoskeleton include DC brushed and brushless motor, Servo motor, Piezoelectric motor and Ultrasonic motor [33].

Recently developed Pneumatic artificial muscle (PAM) contracts like natural muscles and the main advantage is that it offers higher torque to weight ratio. Its impedance is also lower compared to electric motors. Exoskeletons like RUPERT [14] Pneu-Wrex [41], ASSIST [40], Salform arm [49] represent this category. Artificial muscle or Mckibben type of actuators (as shown in Figure 7) are used for pneumatic actuation which has two layers of inner rubber liner and outer periphery made up from braided nylon [65], however, it suffers from a series of problems like lower bandwidth, non-linear characteristics, unidirectional operation and bigger size. Because of the bigger size it is difficult to place in a confined area along with other components. Since it operates in one direction a pair of PAM is required for achieving antagonistic joint motion. Human joint having several degree of freedom such as shoulder joint and wrist joint are difficult to make using PAM.

A few of exoskeleton systems have used hydraulic and pneumatic means as an option for actuation. Because of its size, weight, impedance, leakage and portability problems such systems do not offer any significant advantages other than payload and compliance. System like NEUROEXOS [32] have a big cylinder and pump connected to it therefore it is very difficult to relocate these components during motion. A small flexible fluidic actuator (FFA) has been applied for elbow movement which changes its size depending on the pressure [59] (shown in Figure 7). Systems such as Dampace [37] and Limpact [38] have hydraulic braking which is a product of rotational hydro-elastic actuator. Electro Active Polymer (EAP) is a newly developed elastic material that has many similarities like a human muscle [66]. In this material actuation is generated because of ionic species movement that can be used for micro actuation in exoskeleton (Figure 7). It offers several advantages such as bandwidth and higher levels of electrical-to-mechanical power conversion ratio but has very low torque to weight ratio. For this reason, at present it is not fit for arm exoskeleton actuation.

On the other hand, a passive exoskeleton system may consist of a supporting mechanism to provide variable holding force. For generating the holding force, it uses brake which may be sometimes spring actuated or may have hydraulic or pneumatic drivers. The variable holding/resistive force is required to provide different exercise routines for rehabilitation. Hydraulically driven Dampace is used as passive exoskeleton system [37]. The end effector based system ARM-Guide is based on magnetic particle brake [67], MEM-MRB [24] is magnetorheological fluid based and PLEMO [68] uses electrorheological fluid as the braking technology, the rheological property of those fluids change with the variation of magnetic or electric field. Some attempts have been made to use parallel mechanism for actuation in place of serial mechanism MAHI [52], since parallel manipulators could offer higher stiffness in a confined area. If actuator is placed at the joint the amount of torque required to move the arm is very high whereas if it is connected directly to control the arm it may provide extra mechanical advantage. Some attempts have been made to emulate the actuation mechanism of human arm by using cable based systems which is reported to increases the stiffness by a factor 10 to 12 [69]. Some cable driven exoskeletons have been successfully developed and used for patient rehabilitation such as CADEN-7 [51], SUEFUL [46] MEDARM [30] and Cable driven arm [34, 35].

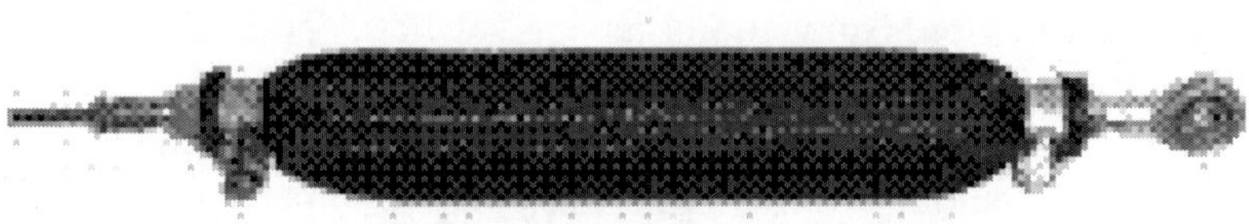

a. Pneumatic artificial muscle [65]

b. Flexible fluidic actuator [59]

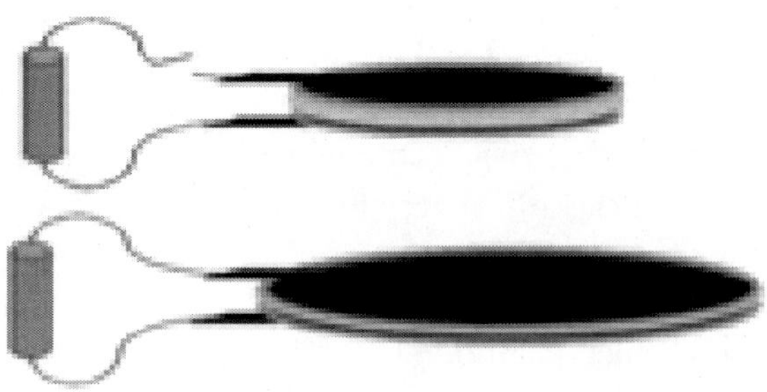

c. Electroactive polymer [66]

Figure 7. New actuation systems.

REDUNDANCY

Redundancy of an exoskeleton is nothing but adding an extra joint to the structure. Though the extra joint adds up complexity to the system, many researchers found that it provides extra advantages as well [42, 46]. Problems such as misalignment, singularity could be reduced with the additional joint. It also helps the system to get the flexibility similar to human arm and increase the level of ergonomic value and comfort [9]. Figure 8 explains this flexibility. Also offset compensation in sternoclavicular joint and wrist joint has been reduced using the redundant joint but at the same time it will create complexity to the design, kinematics and control, though mostly these extra joints are passive in nature.

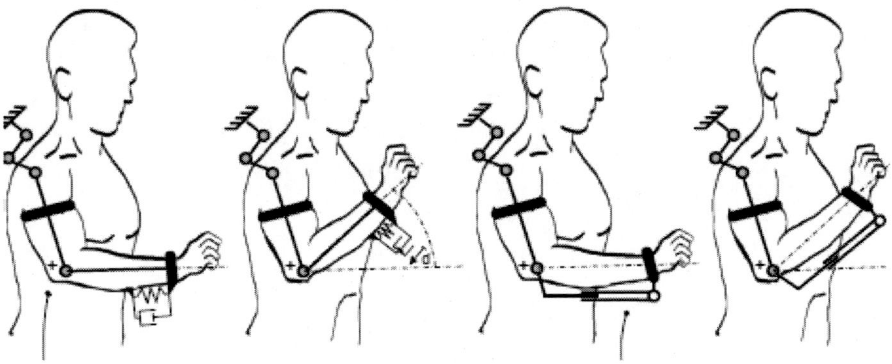

Figure 8. Joint offset compensation using redundant passive DOF [9].

SINGULARITY

When two or more axes attain collinear alignment this results in unpredictable motion due to loss of a degree of freedom. This causes singularity configuration and occurs in a series of revolute joints. In exoskeletons, shoulder internal and external rotation aligned with elbow pronation and supination motion can create such a configuration. Sometimes because of the alignment between human and exoskeleton system may double singularity which will affect both movement and range of motion [63]. In such configurations small linear velocity of arm will require very large joint motions [70]. As a result, motor as well as patient can be affected and will make it difficult to move the arm. From kinematics point of view, the design solution is to avoid or minimise the occurrence of singularity, for example some constraints (joint angle limits) may be applied or some extra conditions could be imposed so that the arm never reaches singularity configurations. Redundant DOF may help overcome the singularity problem [70].

DEGREES OF FREEDOM

Mechanical design of exoskeletons could cover the full human arm, just the elbow joint or allow the full wrist motion. Accordingly, the DOF of exoskeletons could be decided. For a single arm such as ESTEC [50] 9 DOF were allowed, Intelli arm [42] and MIME-RiceWrist [47] had 10 DOF both. A handful of portable exoskeletons, which are commercially available, had 1 DOF such as Hand Mentor [60] and Myomo [61]. Some exoskeletons used passive joints to add a few DOF but are mainly for misalignment compensation, singularity and joint offset reduction. However, it has been shown that rehabilitation process did not depend on the DOF but increased DOF was just to incorporate all the human arm motion in a single system [71, 72].

GRAVITY COMPENSATION

When exoskeleton attempts to lift the human arm against the gravity for ADL and Physical therapy (including its own weight), it is subjected to a varying amount of torque at joints. The actuators have to provide the maximum torque that develops during the course of motion. Larger the torque requirement heavier is the exoskeleton. If a provision is made to offload the actuator toque by compensating the gravity, it will not only improve power requirements of the system but will also make it a portable system. Many exoskeleton designs have used spring actuated link mechanism for gravity compensation [73, 74]. The main criteria in these designs are about selection of the right stiffness of the spring, its material as well as their placement in the system. Some hydraulic based brakes are also used where changing viscous property provides different resistive force during gravity balancing [37].

Coupling

Attachment of exoskeleton to human arm or to soft tissue may create a problem to the patient [50]. Therefore, perfect coupling would be required to place the attachments to the neutral tissue less prone to pain. Shoulder joint is one of main challenges to the designer because of its complexity. Sometimes the centre of rotation of human arm has to be shifted, as shown in Figure 9, with respect to the exoskeleton during the movement [9]. Therefore, the mechanical design has to overcome the ill effects caused by change of centre of rotation of the moving mechanism. Kichugi et al. [75] have designed an exoskeleton for a wheelchair based system with a new mechanism which is able to change its centre of rotation during movement. In ground based exoskeleton systems shoulder joint can be controlled distally resulting in a perfect coupling to any shoulder joint unlike the body based systems. Also axis deviation of wrist flexion/extension and wrist radial/ulnar deviation can be better controlled by having different axes for both wrist motions. In some designs (W-EXOS [55]) the axis offset of wrist joint takes care of this so the undesired joint pain could be avoided.

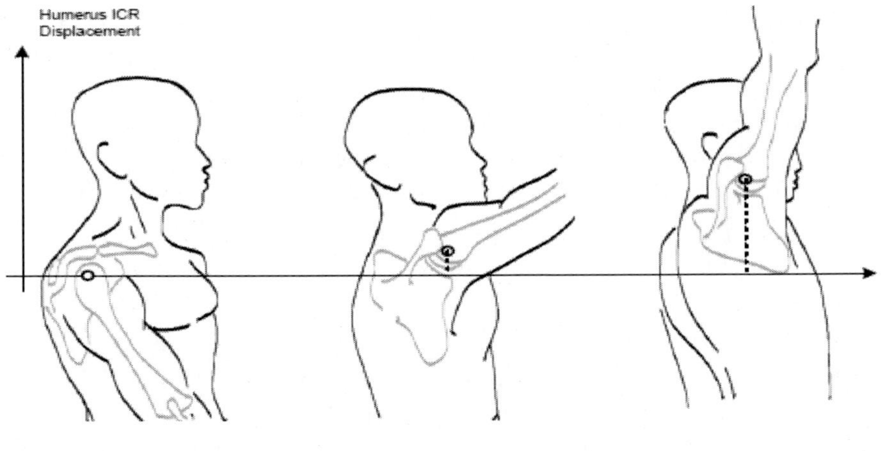

Figure 9. Change of centre of rotation during movement [9].

Bandwidth

Many times exoskeletons are evaluated in terms of their bandwidth i.e., the frequency of the system which also defines the relationship between weight and stiffness [38]. For getting better performance, the lowest natural frequency of the exoskeleton should be higher than the maximum frequency of the human activation signal. This could be achieved by increasing the frequency of the command signal by making amplitude constant to the resonant frequency of the exoskeleton structure. That way it can analyse the effectiveness of exoskeleton system quicker than humans.

EXOSKELETON JOINTS

Human joints are complex and difficult to mechanically replicate in an exoskeleton design. One such joint is shoulder which can be approximated by 3 DOF, elbow appears to be a single joint but in fact it has 2 DOF and in this consideration the wrist joint could be represented by 2 DOF. All human joints are driven by numerous muscle tendon assembles on the bio-electric potential generated at muscles. Putting a limited number of actuators is not sufficient to move the arm in different directions like a natural arm. Several actuators are required in a confined space which is not an easy task to achieve. In serial exoskeleton design joints are connected serially along with supporting linkages, the designer has to make sure that the centre of rotation of actuator matches accurately with the particular joint. The offset between joints should also be least so that it can act like a single joint and is the same for all degrees of freedom. Sometimes end effector based systems are integrated with the exoskeleton to make a hybrid system such as Armeospring [39], BONES [36] and Gentle/G [44]. In such systems the main advantage is that shoulder joint is not fixed to human body therefore it is adaptive to all dimensions and to different patients as shown in Figure 10. System such as HEnRiE [43] is a combined rehabilitation exoskeleton which includes HapticMaster along with two passive joints for grasping.

Sometimes two robots can be used to imitate master-slave operation in therapy; it could be used for haptic operation as well as tele-rehabilitation process. The hybrid exoskeletons are suitable for this purpose, system such as iPAM [57] and UMH [48] fall under this category. ShouldeRO [31] is an alignment-free two DOF rehabilitation robot for the shoulder complex. Parallel manipulators are also proposed for such applications however they have limited stiffness and dexterity. Some exoskeleton systems such as RehabExos [54] have the back drivable facility. It helps the patient to work in passive mode and it allows maximum range of motion to the patient. Pneumatic bending joint using skewed rotary elastic chambers (sREC) [76] are also a soft-actuated system to assist patient in several therapies.

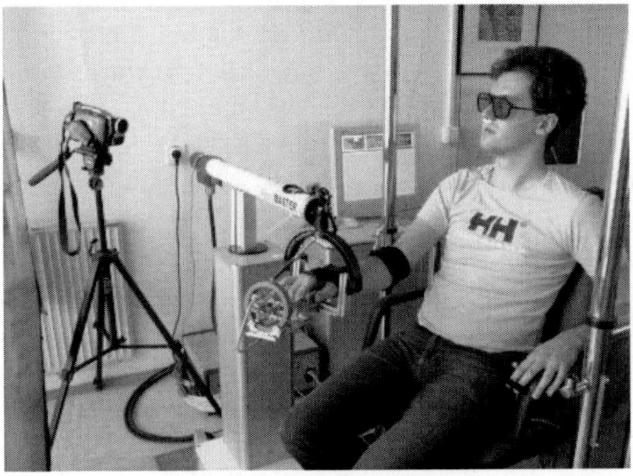

a. Hybrid system [43]

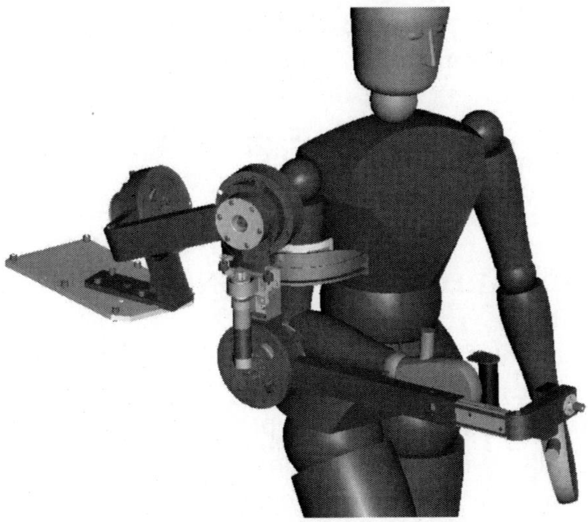

b. Exoskeleton with back drivable facility [54]

Figure 10. Different types of exoskeleton joints.

Ergonomic Challenges

Exoskeleton design has several ergonomic challenges with respect its comforts ability, link variation and softness that the user may be looking for [9]. The design should have a suitable kinematic structure similar to a human arm and the range of motion of exoskeleton should not constrain arm movement. If the structure is made up of rigid linkages the length of supporting link should resemble with human arm. Also the variation in centre of rotation of several joint as well as arm length should be compensated with a flexible design [50]. Material coupled to the human arm should not be discomforting and non-reactive to skin. That is why use of pneumatic actuation has been attempted as wearable jacket in several designs [40, 41, 49, 65]. EMG signals have been used for controlling the exoskeleton system where the sensors extract the signal from arm muscles and through amplification and processing it can be sent to the particular motor to move the arm accordingly. However, the muscle activity is different for different postures of the human arm therefore the position data from different sensors could be also combined with the muscle data to actuate the exoskeleton system. Different ergonomic models have been proposed; pHRI [77] is an interesting design where the real time joint offset and stiffness relationship between human and exoskeleton have been discussed and validated with experimental results.

Modularity and Re-Configurability

Modularity and reconfiguration is the concept where some exoskeletons have been designed by integrating different parts dedicated to a particular need. The parts are detachable and could be assembled as per the requirement. It suits the portability and flexibility of the system and depending on the type of rehabilitation service, the modules could be changed. MUNDUS [19] is a modular system as a result therapy costs are claimed to have been reduced. Also a single system could be used for different movements; Universal Haptic Device is an exoskeleton used for the shoulder and elbow joint can be used for elbow and wrist joint too [78].

FEEDBACK SIGNAL

Initial exoskeletons relied on the position data because it was based on resolved motion rate control or independent joint control [15]. Later PD [41] and PID [49] controls were implemented where velocity and acceleration data are also taken into account along with the position data from sensors for better stabilization and lower steady state error. But since positional data can only analyse the kinematics not the dynamics so in terms of dynamic control forces, torque [47] and inertial sensors [79] have been introduced as the feedback signal. Sometimes an array of sensors is used so that force extraction at human-exoskeleton interface can be monitored during the shifting of exoskeleton with respect to muscle [80]. In impedance [14, 47, 48, 50] and admittance [43, 44] control, kinematic and dynamic data of exoskeleton are merged into the control system to analyse human arm movement with exoskeleton. An EMG (electromyogram) sensor measures the bio-electric potential generated in muscles during the contraction using non-invasive electrodes. This strategy has been used for many exoskeletons for upper-limb motion assist using EMG based control [19, 46, 51, 55, 59, 61, 62, 75]. More recently EEG based Brain machine control [81, 82, 83] has been used to capture the pattern of neural activity in brain that precedes the motion. Voltage fluctuations resulting from neural activity of the brain can be used to actuate the exoskeleton. Also visual [41], audio [41] and haptic [25] signals could be integrated for making a user controlled autonomous system. It is expected that the exoskeleton system will become much more intelligent and sophisticated with advanced controller design and numbers of feedback signals and sensory inputs. This could include capturing motion and physiological data from auxiliary devices such as Mayo Armband [84], Kinect sensor [85] and receiving bio-feedback from EEG or BCI signals, Figure 11 shows a schematic of such a possibility.

PORTABILITY

To use an exoskeleton as body-based assistive device portability is one of the main criteria; ground based exoskeletons are generally suitable for rehabilitation purpose where size of the exoskeleton is not important. For a portable exoskeleton the actuator should be small and less weighty; therefore, the portability of a system should be based on following selection criteria.

I. **Material selection:** Selection of material for building exoskeleton is directly contributing to the weight and hence the portability of the design. Most of the existing designs use metal alloy such as aluminium for constructing the body of exoskeleton. However, with the advancement in new materials which are light as well as strong, portable exoskeletons could be developed using such materials. Materials such as nylon, polypropylene, ABS (Acrylonitrile butadiene styrene) have high strength to weight ration which could be suitable for developing portable and flexible exoskeleton structure.

II. **Energy consumption:** Active exoskeletons require constant power supply for actuators to maintain the joint torque however, all joints do not require the same amount of torque and is also configuration dependent of the exoskeleton arm. Therefore, some joints could be rendered as passive without actuators and the motion could be imparted by releasing the potential energy of springs and some braking mechanism arrangements. In fact, some designs have presented interesting concept of gravity balancing using springs where arm joints are torque balanced at every configuration and only a small motor torque is required to move the arm from statically balanced conditions [73]. This concept is implemented in WREX design [86] and Armon [20]. Since this does not require any energy source and could hold the arm in balanced positions this concept could be implemented to develop a portable system. The passive exoskeletons however do not actuate the joints therefore a hybrid system with optimal combination of active and passive joints should be considered for portable exoskeleton design.

III. **Actuators:** As discussed before actuator selection is one of the main criteria for portability of an exoskeleton because the size and weight are going to directly contribute to this. Most of the existing models are using electromagnetic motor for easier and linear control. However, the size and weight of the motor will vary according to the torque requirement and placement of actuator. Pneumatic and hydraulic actuators require cylinder and pump or compressed gas for actuation so they could form a suitable actuation source but it may be difficult to incorporate the whole actuation system in to a portable device. However, some designs [65] have used pneumatic artificial muscle for joint actuation in a portable system. Some newly developed actuators like Electroactive polymer (EAP) [66] and rheological fluid based systems [38] have the potential but their operational requirements make them unsuitable for portable application. So far with all the existing options used in exoskeleton design electric motors are still the best option in terms of linearity, controllability and size.

IV. **Design:** Ground based exoskeleton system do not have to necessarily consider the size, weight of mechanical construction, the type of material and the use of actuator. When it comes to designing a portable exoskeleton, the designer has to consider every component of the structure and length of the human arm it has to cover so that the joints of exoskeleton and human arm are fully coupled to each other. Also the design should consider system optimization for higher torque to weight ratio for the same structure of the exoskeleton and the actuators used. Spring based mechanism can be used for gravity compensation as well as for providing antagonistic forces which could reduce the size and energy consummation in exoskeleton system.

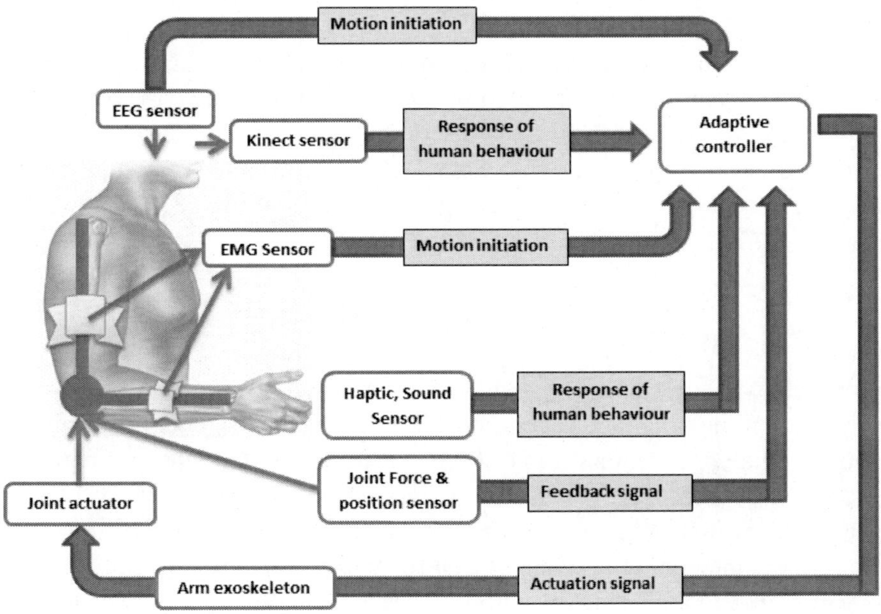

Figure 11. Feedback signal for BCI system.

Hyperextension

Hyperextension is the movement beyond the anatomical limit. It is one of the crucial requirements of designing exoskeletons. Every human joint has its range of motion and beyond that it will cause injury to the joint. The design should be mechanically as well as electronically protected so that it operates in the safe zone [51].

Workspace

The design should be such that it will cover maximum work area in a 3-dimensional space. In exoskeleton design workspace evolution is one of the important criteria since that will allow how effective rehabilitation training could be provided. Majority of exoskeleton have limited workspace because of mechanical restrictions on link length and joints. Higher workspace always adds extra advantage to the ergonomic value of the exoskeleton [52]. This can also avoid singular locations [87, 88] which in turn provide better flexibility in rehabilitation process; it helps visualization of workspace that can relate with the performance evaluation [88]. Several models have been evaluated to increase the workspace such as orthogonal experiment design [87].

Control System Design

Which type of control system should be used in exoskeleton system design is an important consideration. It should take account of different parameters like gravitational force, frictional force and Coriolis components. Initially the approach was focussed on simple position control where mainly positional data from joint sensor were recorded. This was then moved to PID (proportional-integrating-derivative) control where velocity and acceleration data were used to create stability in the exoskeleton system. This was further enhanced by increasing the complexity of the control algorithms with the advent of new sensors and the fast computing power.

When it comes to the dynamic interaction between environment and robotic exoskeleton, force or positon individually is not enough to satisfy the control strategy. Impedance control is based on force feedback where error between the desired position and existing position is computed to create force of end effector during contact. Impedance of a system basically depends upon the task to be performed; it specifies the relationship between interaction force and the desired motion (suited for low contact force). The main disadvantage of impedance control is that accurate force measurements requires compensation for the natural dynamics of the exoskeleton, such as gravity loading and drive friction; it will show poor accuracy in free space due to friction. It can be improved by integrating torque sensor and joint with low friction. L-Exos is a classic example of this kind of exoskeleton [58]. Admittance control is the opposite approach of impedance where the error of joint force is converted into corresponding displacement which has been applied to servo controller to drive the joints to the desired position. Impedance and admittance control are complementary to each other. However, admittance control has the major drawback of instability for dynamic interactions. MGA Exoskeleton [26] is an example of this approach and the more recently ARMin Exoskeleton [12, 22] operate in either admittance or impedance mode. Rahman et al. have proposed a nonlinear sliding mode control technique for controlling the shoulder cum elbow joint and wrist joint [28, 29]. In this approach an exponential reaching law is being used along with the integral of error signal to have better tracking performance. The new trend is to use sophisticated control regimes like neuro-fuzzy, genetic, hybrid and pattern based control by integrating all the bio-signal to make it an intelligent system. Both EMG and EEG based control has been executed with neuro-fuzzy control [46, 55]. It makes synergistic combination of the two techniques; reasoning based fuzzy algorithm with the learning pattern of neural networks. Also in this control some complicacy due to cognitive behaviour of Human can be incorporated [89]. As an example, Fuzzy Hybrid Force-Position Control is used for the robotic arm of an upper limb rehabilitation robot powered by pneumatic muscles [90].

Motion capture system could also be used for incorporating human kinematics directly rather than taking joint data from human coupled sensors because position of sensor changes several times during different postures. The kinematic data of the human arm of a single subject during daily activities were collected using the VICON motion capture system [91] (Figure 12). Kinect camera is also useful for taking 3D data in different postures [85].

However, nowadays the above control strategies are taking back stage and adaptive control is increasingly becoming popular for exoskeleton control since it can help change the mode of exercises, level of difficulty and force level depending on the patient's requirement. This can also include other forms of data like visual, audio and emotions to make ultimate determination of the control action. There should also be a provision of manual override together with local intelligence system for safety purposes. The overall system could be a hybrid control system where master/slave as well as automatic local control scheme may be integrated for rehabilitation and if there is any problem in the local control system due to malfunction, it should be controlled by the master controller from outside.

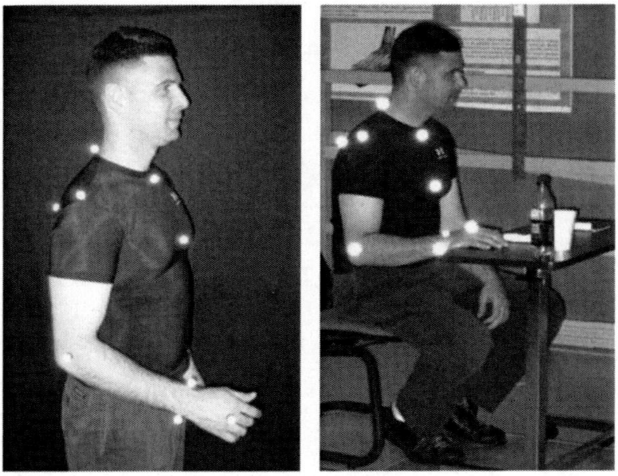

Figure 12. VICON motion capture system [91].

REHABILITATION PROCESS

Exoskeletons are being evaluated in terms of structural efficacy; however, recently such systems are subjected to clinical trial for measuring their effectiveness for clinical use. Clinical requirement of rehabilitation will depend upon different grounds such as psychological, medical and ergonomic.

The task oriented motor learning program improves motor functions. Several new innovative techniques of therapy like constrained induced movement therapy, electro-myographical bio- feedback based are also implemented under this scheme [92]. To improve the motor function in a better way, patient engagement is required during the exercises. It is patient's personal effort which will help to make the whole process effective and subjective. Such task based rehabilitation exercise will motivate the patient as they can engage with the game in a virtual platform which may resemble with the kind of movement required for the therapy. The difficulty level can be varied in the game according to the level of exercises. This will create an interesting platform and a more entertaining therapy process which will make the patient to take part positively. An example of this is the development of VR platform for stroke neuro-rehabilitation called the Rehabilitation Gaming System (RGS) [93]. It is based on goal-oriented movement which improves motor neuron system. It is not ascertained how much patient could withstand in terms of intensity of the training, it is yet to be seen what intensity dose of training is the right one for individual patient and for each stage of the rehabilitation process. Milot et al. [71] has examined a comparative study using one exoskeleton having single DOF with another multi DOF exoskeleton system. But surprisingly after a week of exercises, the net result in terms of improvement in muscle function are the same in both cases. So this provides some insight that a simpler system could be as effective as a complex one if rehabilitation is planned properly.

Also it needs to be investigated whether bilateral training or unilateral training are more beneficial for the patient. Experiments in post stroke therapy using progressive task specific in two modes-unilateral and bilateral have shown that bilateral training are more useful for the patient [94]. It is yet to be seen how stroke patients will respond to the robotic therapy after 6 months of rehabilitation. However, it's quite interesting to see that exoskeleton based training can play a significant role for rehabilitation process [22, 94, 95]. This invites some crucial question:

i. How should the difficulty level of the therapy and its goal be tuned to optimize the patient engagement?
ii. Is it possible to adjust the difficulty level in real-time to promote the active participation of the patient?

TELE REHABILITATION

In tele-robotic rehabilitation system, a master arm could remotely control an exoskeleton far away from the place of operation [96]. Such systems can provide guided training supported by experts. RehaBot is an Advanced Rehabilitation Robotic System working on this principle for Augmenting Healthcare [27]. It provides real time data and force feedbacks to the therapist to execute training for the patient situated at a remote location with no delay in data transmission [96]. The schematic shown in Figure 13 could be a useful training regime for providing rehabilitation supported by expert guidance.

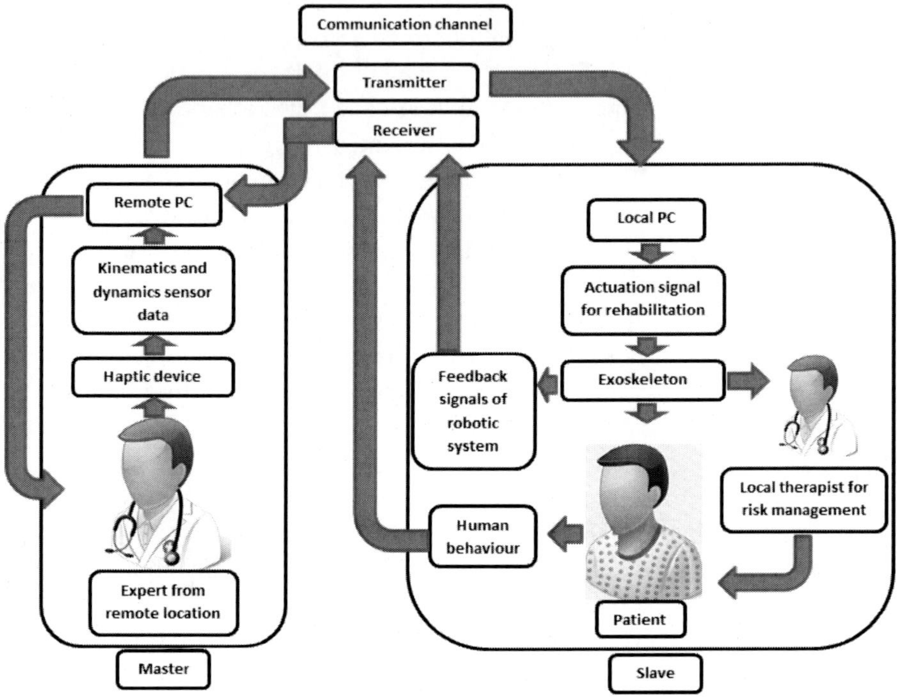

Figure 13. Tele-rehabilitation architecture.

Clinical Scale for Rehabilitation Evaluation

Several scales have been proposed for evaluating effectiveness of rehabilitation to measure the patient condition after stroke [71, 95]. However most of the clinical scales are being used for assessment of sensory and motor functions. Many organizations have made their own scale of assessment such as Nottingham sensory assessment [71], California Functional Independence Scale [94], Modified Ashworth scale [71]. But after thorough review, mostly two scales seem to be used for evaluation of the recovery of the patient suffering from different neuromuscular diseases. Those are:

1. Fugl-Meyer Motor scale (FMA)
2. Wolf-Motor function test (WMFT)

FMA scale is not only used for stroke assessment but for other activities such as ADL, Functional Mobility and pain. It includes all the probable measurable parameters of human body for assessment of post rehabilitation condition such as motor function, sensory function, balance, joint range of motion and joint pain. On the other hand, WMFT is applicable to stroke, dexterity, strength and upper extremity function. It basically considers task based rehabilitation therapy in terms of time, functional ability and strength. So far all the scales developed are used for manual therapy process. However, till now there is no standard scale especially defined to measure the improvement after rehabilitation using exoskeleton. So there is a need to develop some standards to analyse the clinical effectiveness of exoskeleton. Most research papers have reported the study of mechanical advantage of the exoskeletons, however, some recent are trying to relate the rehabilitation outcomes with fMRI of brain; if any improvement in the neuro motor function happens it will show up in the lesion portion of brain through fMRI [97] as shown in Figure 14.

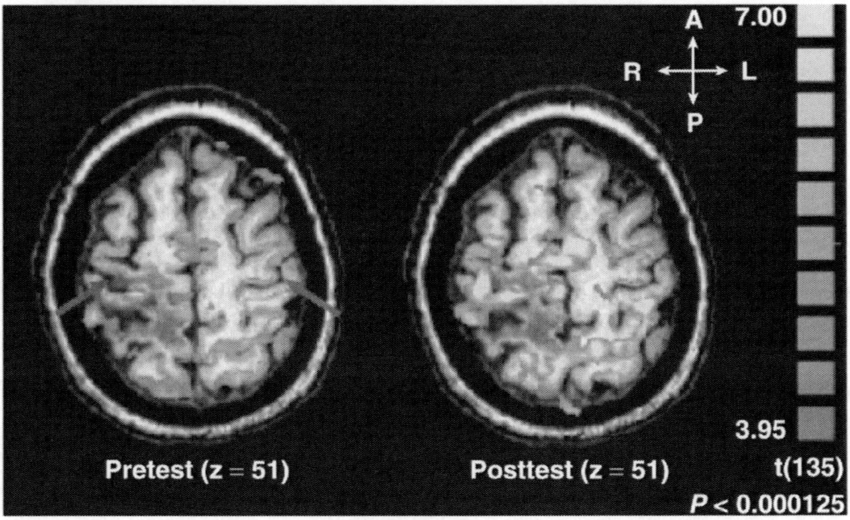

Figure 14. Pre and post rehabilitation brain fMRI [97].

Handy Electronic Gazette

With the advancement in technology new and inexpensive electronic products such as Mayo Armband [84], Google glasses [98], Eye Tracking systems [99] and Kinect [85] are available to be integrated with the exoskeleton systems. Those systems can provide auxiliary support either for controlling the device or for receiving feedback from the devices. Currently some exoskeletons are being controlled by taking the EMG signals from muscles, with the Armband this could be easily achieved from the modular design of the device [84]. The Kinect camera could be used for 3D measurement as well as emotional evaluation of the patient during rehabilitation training. This could be also used for game based training where Kinect sensor can be used for pattern recognition and object tracking; such examples are available for use with exoskeleton [85]. A few other new gadgets like Google glass and eye tracking system can be incorporated with the exoskeleton system for live interaction as well as for user interface with eye tracking coordination for special needs patients.

Conclusion

It is evident from this review that there still exist lots of opportunities in design and development of upper arm exoskeletons. These are in structural design of the exoskeleton including materials, mechanism for exoskeleton and actuators (design and selection), control strategy and user interface. The actuator should provide higher torque to weight ratio and the mechanism should offer higher flexibility with less number of actuators. It should be possible to design a simple but intelligent mechanism which will offer enhanced torque to weight ratio without affecting its working principle. The exoskeleton may have minimum number of active joints so that it will be wearable by human users. Provision of passive joint will compensate misalignment and offer better coupling with arm which should be considered in the design.

Most of the systems are based on impedance or admittance control where force and position related data is used as a feedback signal, however since the exoskeleton is going to be used by humans the control should be based on the bio-signals coming from human body. This could include taking EMG, EEG for activation and mechanical joint data for control. This could be further integrated with new electronic gadgets like Armband and external sensors for user specific control such as by incorporating audio, visual, haptic and tactile feedback.

Energy consumption is a big issue particularly when the system is to be portable. The control system should be adaptive in nature which can change its characteristics depending on the task requirement i.e., if the training routine is known in advance the actuators should operate optimally for energy conservation. Also depending on the patient requirement and recovery, the controller should adopt the level of difficulty on exercises. There is another area which seems to be completely untouched is about standardisation of arm rehabilitation using robotic exoskeletons so that patients could use a portable arm rehabilitator and select a program to suit their personal requirements which can not only train with real time patient parameters but also evaluate their recovery status as well as providing assistance in activities of their daily living. Once these requirements are full filled there is no doubt such exoskeletons will get a wider user acceptability.

REFERENCES

[1] Maciejasz, P; Eschweiler, J; Hahn, KG; Troy, AJ; and Leonhardt, S. A survey on robotic devices for upper limb Rehabilitation. *Journal of Neuroengineering and Rehabilitation*, 9th January 2014, http://www.jneuroengrehab.com/content/11/1/3.

[2] Chang, WH; Kim, YH. Robot-assisted Therapy in Stroke Rehabilitation. *Journal of Stroke,* September 2013, 15(3), 174-181.

[3] Stroke Association. A complete guide to cognitive problems after stroke. April 2015. URL: https://www.stroke.org.uk.

[4] Stroke Association. Aids and equipment for independent living. June 2015. URL: https://www.stroke.org.uk.

[5] Krakauer, JW; Marshall, RS. The Proportional Recovery Rule for Stroke Revisited. *Annals of Neurology*, December 2015, 78(6), 847-847.

[6] Lo, SH; Xie, SQ. Exoskeleton robots for upper-limb rehabilitation: State of the art and future prospects. *Medical Engineering and Physics*, April 2012, 34(3), 261-268.

[7] Krebs, HI; Hogan, N; Aisen, ML; and Volpe, BT. Robot-aided neuro-rehabilitation. *IEEE Transaction on Rehabilitation Engineering*, March 1998, 6 (1), 75–87.

[8] Spencer, SJ; Klein, J; Minakata, K; Le, V; Bobrow, JE; and Reinkensmeyer, DJ. A low cost parallel robot and trajectory optimization method for wrist and forearm rehabilitation using the Wii. *Proc. 2nd IEEE RAS and EMBS International Conference on Biomedical Robotics and Biomechatronics (BioRob)*, Scottsdale: AZ, 19th – 22nd October 2008, 869–874.

[9] Schiele, A. Kinematic Compatibility between Human Limbs and Exoskeleton Structures in Fundamentals of Ergonomic Exoskeleton Robots. In: Schiele, A. *Wearable Robots: Biomechatronic Exoskeletons*. Duitsland: John Wiley and Son; 2008; 19-22.

[10] Umphred, DA; Lazaro, RT; Roller, M; and Burton, G. Neurological Rehabilitation, 6th edition. Elsevier Health Sciences; 25th September 2012.

[11] Dijkers, MP; DeBear, PC; Erlandson, RF; Kristy, K; Geer, DM; and Nichols, A. Patient and staff acceptance of robotic technology in occupational therapy: a pilot study. *Journal of rehabilitation Research and Development*, 1991, 28(2), 33–44.

[12] Nef, T; Mihelj, M; Kiefer, G; Pemdl, C; Muller, R; and Riener, R. ARMin Exoskeleton for Arm Therapy in Stroke Patients. *Proceedings of the 2007 IEEE 10th International Conference on Rehabilitation Robotics,* Noordwijk: Netherlands, 12th -15th June 2007, 68-74.

[13] Mistry, M; Mohajerian, P; and Schaal, S. Arm Movement Experiments with Joint Space Force Fields using an Exoskeleton Robot by. *Proceedings of the 2005 IEEE 9th International Conference on Rehabilitation Robotics,* Chicago, IL: US, 28th June – 1st July, 2005, 408-413.

[14] Zhang, H; Austin, H; Buchanan, S; Herman, R; Koeneman, J; and He, J. Feasibility Studies of Robot-Assisted Stroke Rehabilitation at Clinic and Home Settings Using RUPERT. *2011 IEEE International Conference on Rehabilitation Robotics Rehab Week,* Zurich: Switzerland, 29th June – 1st July 2011, 1-6.

[15] Manna, SK; Bhaumik, S. A Bio-inspired 10 DOF wearable powered arm exoskeleton for rehabilitation. *Journal of Robotics: Rehabilitation Robotics,* 8th October 2013, 2013.

[16] Sankai, Y. HAL: Hybrid Assistive Limb based on Cybernics. *Springer Tracts in Advanced Robotics,* 2011, 66, 25-34.

[17] Rocon, E; Belda-Lois, JM; Ruiz, AF; Manto, M; Moreno, JC; and Pons, JL. Design and validation of a rehabilitation robotic exoskeleton for tremor assessment and suppression. *IEEE Transaction of Neural System and rehabilitation Engineering*, September 2007, 15(3), 367–378.

[18] Loureiro, RCV; Belda-Lois, JM; Lima, ER; Pons, JL; Sanchez-Lacuesta, JJ; and Harwin, WS. Upper limb tremor suppression in ADL via an orthosis incorporating a controllable double viscous beam actuator. *Proc. 9th Int. Conf. on Rehabilitation Robotics ICORR,* Chicago, IL: USA, 28th June-1st July 2005, 119- 122.

[19] Pedrocchi, A; Ferrante, S; Ambrosini, E; Gandolla, M; Casellato, C; Schauer, T; Klauer, C; Pascual, J; Vidaurre, C; Gfoehler, M; Reichenfelser, W; Karner, J; Micera, S; Crema, A; Molteni, F; Rossini, M; Palumbo, G; Guanziroli, E; Jedlitschka, A; Hack, M; Bulgheroni, M; D Amico, E; Schenk, P; Zwicker, S; Duschau-Wicke, A; Miseikis, J; Graber, L; and Ferrigno, G. MUNDUS project: Multimodal Neuroprosthesis for daily upper limb support. *Journal of Neuroengineering and Rehabilitation*, 3rd July 2013, http://www.jneuroengrehab.com/ content/10/ 1/66.

[20] Herder, JL; Vrijlandt, N; Antonides, T; Cloosterman, M; and Mastenbroek, PL. Principle and design of a mobile arm support for people with muscular weakness *Journal of Rehabilitation Research and development,* August/September 2006, 43(5), 591-604.

[21] Johnson, GR; Carus, DA; Parrini, G; Marchese, SS; and Valeggi, R. The design of a five-degree-of-freedom powered orthosis for the upper limb. *Proc Inst Mech Eng H,* 2001, 215(3), 275–284.

[22] Marganska, VK; Blanco, J; Campen, K; Curt, A; Dietz, V; Ettlin, T; Felder, M; Fellinghauer, B; Guidali, M; Kollmar, A; Luft, A; Nef, T; Amft, CS; Stahel, W; and Riener, R. Three-dimensional, task-specific robot therapy of the arm after stroke: a multicentre, parallel-group randomised trial, *Lancet Neural*, February 2014, 13(2), 159–166.

[23] Manna, SK; Bhaumik, S. Hand Exercising Exoskeleton: An Aging Assistive Device for Rehabilitation. In: Morsi, YS., Shukla, A. and Rathore, CP. *Optimizing Assistive Technologies for Aging Populations.*PA, USA: IGI Global; 2015; 78-96.

[24] Oda, K; Isozumi, S; Ohyama, Y; Tamida, K; Kikuchi, T; and Furusho, J. Development of isokinetic and iso-contractile exercise machine MEM-MRB using MR brake. *Proc. IEEE Int. Conf. on Rehabilitation Robotics (ICORR)*, Kyoto: Japan, 23rd -26th June 2009, 6–11.

[25] Letier, P; Roosevelt, FD; Motard, E; and Verschueren, JP. EXOSTATION: Haptic exoskeleton based control station. *Proceedings - IEEE International Conference on Robotics and Automation*, Anchorage: AK, 3rd -7th May 2010, 1840 – 1845.

[26] Carignan, C; Tang, J; Roderick, S; and Naylor, M. A Configuration-Space Approach to Controlling a Rehabilitation Arm Exoskeleton. *Proc. IEEE 10th International Conference on Rehabilitation Robotics ICORR*, Noordwijk: Netherlands, 13th -15th June 2007, 179 - 187.

[27] Hu, J; Lim, YJ; Ding, Y; Paluska, D; Solochek, A; Laffery, D; Bonato, O; and Marchessaul, R. An Advanced Rehabilitation Robotic System for

Augmenting Healthcare. *33rd Annual International Conference of the IEEE EMBS,* Boston, Massachusetts: USA, 30th August – 3rd September 2011, 2073-2076.

[28] Rahman, MH; Ouimet, TK; Saad, M; Kenné, JP; and Archambault, PS. Development and Control of a Wearable Robot for Rehabilitation of Elbow and Shoulder Joint Movements. *IECON 2010 - 36th Annual Conference on IEEE Industrial Electronics Society*, Glendale, AZ: USA, 07th – 10th November 2010, 1506-1511.

[29] Rahman, MH; Ouimet, TK; Saad, M; Kenné, JP; and Archambault, PS. Modelling and Development of an Exoskeleton Robot for Rehabilitation of Wrist Movements, *2010 IEEE/ASME International Conference on Advanced Intelligent Mechatronics,* Montréal: Canada, 6th -9th July 2010, 25-30.

[30] Ball, SJ; Brown, IE; and Scott, SH. MEDARM: a rehabilitation robot with 5DOF at the shoulder complex. *2007 IEEE/ASME International Conference on Advanced Intelligent Mechatronics*, Zurich: Switzerland, 4th -7th Sept. 2007, 1 – 6.

[31] Dehez, B; Sapin, J. ShouldeRO, an alignment-free two-DOF rehabilitation robot for the shoulder complex. *2011 IEEE International Conference on Rehabilitation Robotics Rehab Week*, Zurich: Switzerland, 29th June – 1st July 2011.

[32] Sardellitti, I; Cattin, E; Vecchi, SRF; Carrozza, MC; Dario, P; Artemiadis, PK; and Kyriakopoulos, KJ. Description, Characterization and Assessment of a Bio- Inspired Shoulder Joint-First Link Robot for Neuro- Robotic Application. *The First IEEE/RAS-EMBS International Conference on Biomedical Robotics and Biomechatronics*, Pisa: Italy, 20th -22nd February 2006, 112 – 117.

[33] Jang, M; Bailak, G. Control System for Multiple Joint Robotic Arm Powered by Ultrasonic Motor. *Applied Power Electronics Conference and Exposition,* Anaheim, CA: USA, 22nd -26th February 2004, 1844 – 1848.

[34] Agrawal, SK; Dubey, VN; Gangloff. J; and Sangwan, V. Design and Optimization of a Cable Driven Upper Arm Exoskeleton, *Journal of Medical Devices*, 31st August 2009, 3(3), 0310004 (1-8).

[35] Brackbill, EA; Mao, Y; Agrawal, SK; Annapragada, M; and Dubey, VN. Dynamics and Control of a 4-dof Wearable Cable-driven Upper Arm Exoskeleton. *IEEE International Conference on Robotics and Automation,* Kobe: Japan, 12th -17th May 2009, 2300 – 2305.

[36] Klein, J; Spencer, S; Allington, J; Bobrow, JE; and Reinkensmeyer, DJ. Optimization of a parallel shoulder mechanism to achieve a high-force, low-mass, robotic-arm exoskeleton. IEEE Transactions on Robotics, 28th June 2010, 26(4), 710–715.

[37] Stienen, A; Hekman, E; Prange, G; Jannink, M; Aalsma, A; van der Helm, F; and van der Kooij, H. Dampace: Design of an exoskeleton for force-coordination training in upper-extremity rehabilitation. *J Med Devices,* 31st August 2009, 3(3), 031003(1-10).

[38] Stienen, AHA; Hekman, EEG; Ter Braak, H; Aalsma, AMM; Van der Helm, FCT; and Van der Kooij, H. Design of a rotational hydro-elastic actuator for an active upper-extremity rehabilitation exoskeleton. *Proc. 2nd IEEE RAS and EMBS International Conference on Biomedical Robotics and Biomechatronics (BioRob)*, Scottsdale: AZ, 19th -22nd October 2008, 881–888.

[39] Sanchez, R; Reinkensmeyer, D; Shah, P; Liu, J; Rao, S; Smith, R; Cramer, S; Rahman, T; and Bobrow, J. Monitoring functional arm movement for home-based therapy after stroke. *Proc IEEE EngMed Biol Soc*, San Francisco: CA, 1st -5th September 2004, 4787 – 4790.

[40] Sasaki, D; Noritsugu, T; and Takaiwa, M. Development of Active Support Splint driven by Pneumatic Soft Actuator (ASSIST). *Proc. IEEE International Conference on Robotics and Automation (ICRA)*, Barcelona: Spain, 18th -22nd April 2005, 520 – 525.

[41] Secoli, R; Milot, MH; Rosati, G; and Reinkensmeyer, DJ. Effect of visual distraction and auditory feedback on patient effort during robot-assisted movement training after stroke. *Journal of NeuroEngineering and Rehabilitation*, 23rd April 2011, http://www.jneuroengrehab.com/content/8/1/21.

[42] Ren, Y; Park, HS; and Zhang, LQ. Developing a whole-arm exoskeleton robot with hand opening and closing mechanism for upper limb stroke rehabilitation. Proc. IEEE International Conference on Rehabilitation Robotics (ICORR). Kyoto: Japan, 23rd-26th June 2009, 761–765.

[43] Mihelj, M; Podobnik, J; and Munih, M. HEnRiE - Haptic environment for reaching and grasping exercise. *Proc. 2nd IEEE RAS and EMBS International Conference on Biomedical Robotics and Biomechatronics (BioRob).* Scottsdale: AZ, 19th -22nd October 2008, 907–912.

[44] Loureiro, RCV; Harwin, WS. Reach and Grasp Therapy: Design and Control of a 9-DOF Robotic Neuro-rehabilitation System. *Proc. IEEE 10th International Conference on Rehabilitation Robotics (ICORR)*, Noordwijk: Netherlands, 13th-15th June 2007, 757–763.

[45] Mayr, A; Kofler, M; and Saltuari, L. ARMOR: an electromechanical robot for upper limb training following stroke. A prospective randomized controlled pilot study. *Handchir Mikrochir Plast Chir*, February 2008, 40 (1), 66–73.

[46] Gopura, RARC; Kiguchi, K; and Li, Y. SUEFUL-7: A 7DOF upper-limb exoskeleton robot with muscle-model-oriented EMG-based control. *IEEE/RSJ International Conference on Intelligent Robots and Systems (IROS)*, St. Louis: MO, 10th -15th October 2009, 1126–1131.

[47] Gupta, A; O'Malley, M; Patoglu, V; and Burgar, C. Design, control and performance of RiceWrist: a force feedback wrist exoskeleton for rehabilitation and training. *Int J Robot Res,* February 2008, 27(2), 233 - 251.

[48] Morales, R; Badesa, FJ; García-Aracil, N; and Sabater, JM. Pneumatic robotic systems for upper limb rehabilitation. *Med Biol Eng Comput*, October 2011 49(10), 1145–1156.

[49] Tsagarakis, N; Caldwell, D. Development and control of a soft-actuated exoskeleton for use in physiotherapy and training. *Autonomous Robots,* July 2003, 15(1), 21–33.

[50] Schiele, A; Van der Helm, FCT. Kinematic Design to Improve Ergonomics in Human Machine Interaction. *IEEE Transactions on Neural Systems and Rehabilitation Engineering*, December 2006, 14(4), 456–469.

[51] Perry, JC; Rosen, J; and Burns, S. Upper-limb powered exoskeleton design. *IEEE/ASME Transactions on Mechatronics*, August 2007, 12(4), 408–417.

[52] French, JA; Rose, CG; and O'Malley, MK. System Characterization of MAHI EXO-II: A Robotic Exoskeleton for Upper Extremity Rehabilitation *Proc ASME Dyn Syst Control Conference,* San Antonio, TX: USA, 22nd-24th October 2014, 1-5.

[53] Housman, SJ; Le, V; Rahman, T; Sanchez, RJ; and Reinkensmeyer, DJ. Arm- Training with T-WREX after Chronic Stroke: Preliminary Results of a Randomized Controlled Trial. *Proc. IEEE 10th International Conference on Rehabilitation Robotics ICORR,* Noordwijk: Netherlands, 13th-15th June 2007, 562–568.

[54] Vertechy, R; Frisoli, A; Dettori, A; Solazzi, M; and Bergamasco, M. Development of a new exoskeleton for upper limb rehabilitation. *Proc. IEEE International Conference on Rehabilitation Robotics (ICORR)*, 23rd-26th June 2009, Kyoto: Japan, 188–193.

[55] Gopura, RAR; Kiguchi, K. A human forearm and wrist motion assist exoskeleton robot with EMG-based Fuzzy-neuro control. *Proc. 2nd IEEE RAS and EMBS International Conference on Biomedical Robotics and Biomechatronics (BioRob),* 19th-22nd October 2008, Scottsdale: AZ, 550–555.

[56] Pignolo, L; Dolce, G; Basta, G; Lucca, LF; Serra, S; and Sannita, WG. Upper limb rehabilitation after stroke. ARAMIS a "robo-mechatronic" innovative approach and prototype. *4th IEEE RAS and EMBS Int. Conf. Biomedical Robotics and Biomechatronics (BioRob),* 24th-27th June 2012, Rome: Italy, 1410–1414.

[57] Culmer, PR; Jackson, AE; Makower, SG; Cozens, JA; Levesley, MC; Mon-Williams, M; and Bhakta, B. A novel robotic system for quantifying arm kinematics and kinetics: description and evaluation in therapist-assisted passive arm movements post-stroke. *J Neurosci Methods*, 30th April 2011, 197(2), 259–269.

[58] Frisoli, A; Procopio, C; Chisari, C; Creatini, I; Bonfiglio, L; and Bergamasco, M. Positive effects of robotic exoskeleton training of upper limb reaching movements after stroke, *Journal of NeuroEngineering and Rehabilitation*, 9th June 2012, http://www.jneuroengrehab.com/content/9/1/36.

[59] Pylatiuk, C; Kargov, A; Gaiser, I; Werner, T; Schulz, S; and Bretthauer, G. Design of a flexible fluidic actuation system for a hybrid elbow orthosis. *Proc. IEEE International Conference on Rehabilitation Robotics (ICORR),* 23rd-26th June 2009, Kyoto: Japan, 167–171.

[60] Koeneman, EJ; Schultz, RS; Wolf, SL; Herring, DE; and Koeneman, JB. A pneumatic muscle hand therapy device. *2nd IEEE RAS and EMBS International Conference on Biomedical Robotics and Biomechatronics*, 19th-22nd October 2008, Scottsdale: AZ, 919 - 923.

[61] Stein, J; Narendran, K; McBean, J; Krebs, K; and Hughes, R. Electromyography-controlled exoskeletal upper-limb-powered orthosis for exercise training after stroke. *American Journal of Physical Medicine and Rehabilitation,* April 2007, 86(4), 255–261.

[62] Pirondini, E; Coscia, M; Marcheschi, S; Roas, G; Salsedo, F; Frisoli, A; Bergamasco, M; and Micera, S. Evaluation of the effects of the Arm Light Exoskeleton on movement execution and muscle activities: a pilot study on healthy Subjects. *Journal of NeuroEngineering and Rehabilitation,* 23rd January 2016.

[63] Malosio, M; Pedrocchi, BN; Vicentini, F; and Tosatti, LM. Analysis of elbow-joints misalignment in upper-limb exoskeleton. *2011 IEEE*

International Conference on Rehabilitation Robotics, 29th June -1st July 2011, Zurich: Switzerland, 1 - 6.

[64] Jansen, J; Richardson, B; Pin, F; Lind, R; and Birdwell, J. Exoskeleton for Soldier Enhancement Systems Feasibility Study. Tennessee: OAK RIDGE NATIONAL LABORATORY; September 2000.

[65] Kobayash, H. Development of a Muscle Suit for Realizing All Motion of the Upper Limb. *Proceedings of 2004 IEEORSJ International Conference on intelligent Robots and Systems*, 28th September- 2nd October 2004, Sendai: Japan, 1630 – 1635.

[66] Hansona, D; Pioggiab, G; Cohenc, YB; and Androids, DDR. Application of EAP as artificial muscles to entertainment industry, *Proceedings of EAPAD, SPIE's 8th Annual International Symposium on Smart Structures and Materials*, 5th -8th March 2001, Newport: CA, 4329-74.

[67] Reinkensmeyer, DJ; Kahn, LE; Averbuch, M; McKenna-Cole, A; Schmit, BD; and Rymer, WZ. Understanding and treating arm movement impairment after chronic brain injury: progress with the ARM guide. *J Rehabil Res Dev,* November-December 2000, 37(6), 653–662.

[68] Kikuchi, T; Xinghao, H; Fukushima, K; Oda, K; Furusho, J; and Inoue, A. Quasi-3-DOF rehabilitation system for upper limbs: its force-feedback mechanism and software for rehabilitation. *Proc. IEEE 10th International Conference on Rehabilitation Robotics (ICORR)*, Noordwijk: Netherlands, 13th-15th June 2007, 24–27.

[69] Ball, SJ; Brown, IE; and Scott, SH. A planar 3DOF robotic exoskeleton for rehabilitation and assessment. *Proceedings of the 29th Annual International Conference of the IEEE EMBS, Lyon*: France, 22nd-26th August 2007, 4024 – 4027.

[70] Lu, J; Chen, W; and Tomizuka, M. Kinematic Design and Analysis of a 6-DOF Upper Limb Exoskeleton Model for a Brain-Machine Interface Study. *Proceedings of the 6th IFAC Symposium on Mechatronic Systems*, Hangzhou: China, 10th-12th April 2013, 293-300.

[71] Milot, MH; Spencer, SJ; Chan, V; Allington, JP; Klein, J; Chou, C; Bobrow, JE; Cramer, SC; and Reinkensmeyer, DJ. A crossover pilot study evaluating the functional outcomes of two different types of robotic movement training in chronic stroke survivors using the arm exoskeleton BONES. *Journal of NeuroEngineering and Rehabilitation*, 19th December 2013, http://www.jneuroengrehab.com/content/10/1/ 112.

[72] Grimaldi, G; Manto, M. Functional impacts of exoskeleton-based rehabilitation in chronic stroke: multi-joint versus single-joint robotic

training. *Journal of NeuroEngineering and Rehabilitation,* 19th December 2013, http://www.jneuroengrehab.com/content/10/1/113.

[73] Dubey, VN; Agrawal, SK. Study of an upper arm exoskeleton for gravity balancing and minimization of transmitted forces, *Proc Inst Mech Eng H,* November 2011, 225(11), 1025-1035.

[74] Chen, Y; Fan, J; Zhu, Y; Zhao, J; and Cai, H. A passively safe cable driven upper limb rehabilitation exoskeleton, *Technology and Health Care*, 17th June 2015, 23 (s2) S197–S202.

[75] Kiguchi, K. Active Exoskeletons for Upper-Limb Motion Assist, *J. Humanoid Robotics,* September 2007, 4(3), 607-624.

[76] Wilkening, A; Stoppler, H; and Ivlev, O. Adaptive Assistive Control of a Soft Elbow Trainer with Self-Alignment using Pneumatic Bending Joint, *2015 IEEE International Conference on Rehabilitation Robotics (ICORR),* 11th -14th August 2015, Singapore, 729 – 734.

[77] Schiele, A. An Explicit Model to Predict and Interpret Constraint Force Creation in pHRI with Exoskeletons, *IEEE International Conference on Robotics and Automation*, 19th -23rd May 2008, Pasadena: CA, 1324 - 1330.

[78] Oblak, J; Cikajlo, I; and Matjacic, Z. Universal Haptic Drive: A Robot for Arm andWrist Rehabilitation. *IEEE Trans Neural Syst Rehabil Eng,* 20th October 2009, 293 – 302.

[79] Song, Z; Guo, S; Xiao, N; Gao, B; and Shi, L. Implementation of Human-Machine Synchronization Control for Active Rehabilitation Using an Inertia Sensor, *Sensors,* 22nd September 2012, 12, 16046-16059.

[80] Duque, JT; Ugalde, RC; Kilicarslan, A; Venkatakrishnan, A; Soto, R; and Contreras-Vidal, JL. Real-Time Strap Pressure Sensor System for Powered Exoskeletons, *Sensors,* 16th February 2015, 15 (2), 4550-4563.

[81] Viriyasaksathian, B; Khemmachotikun, S; Kaimuk, DPP; and Wongsawat, Y. EEG Feedback Training for Upper-Limb Rehabilitation. *Proceedings of the 2011 IEEE International Conference on Robotics and Biomimetic,* 7th-11th December 2011, Phuket: Thailand, 1567 - 1572.

[82] Blank, A; O'Malley, MK; Francisco, GE; and Contreras-Vidal, JL. A Pre-Clinical Framework for Neural Control of a Therapeutic Upper-Limb Exoskeleton. *Int IEEE EMBS Conf Neural Eng.*, 6th-8th November 2013, San Diego, CA: USA, 1159–1162.

[83] Lebedev, MA; Nicolelis, MA. Brain–machine interfaces: past, present and future, *Trends in Neurosciences,* September 2006, 29 (9), 536-546.

[84] Nymoen, k; Haugen, MR; and Jensenius, AR. MuMYO — Evaluating and Exploring the MYO Armband for Musical Interaction, *Proceedings of the International Conference on New Interfaces for Musical Expression*, 31st May- 3rd June 2015, LA, USA, 215-218.

[85] Guevara, DC; Vietri, G; Prabakar, M; and Kim, JH. Robotic Exoskeleton System Controlled by Kinect and Haptic Sensors for Physical Therapy, *29th Southern Biomedical Engineering Conference (SBEC)*, 3rd -5th May 2013, Miami: FL, 71-72.

[86] Rahman, T; Sample, W; Jayakumar, S; King, MM; Wee, JY; Seliktar, R; Alexander, M; Scavina, M; and Clark, A. Passive exoskeletons for assisting limb movement. *Journal of Rehabilitation Research and Development,* August-September 2006, 43(5), 583–590.

[87] Chen, Y; Zhang, J; Yang, C; and Niu, B. The workspace mapping with deficient-DOF space for the PUMA 560 robot and its exoskeleton arm by using orthogonal experiment design method, *Robotics and Computer-Integrated Manufacturing*, August 2007, 23 (4), 478–487.

[88] Gunasekara, MP; Gopura, RARC., Workspace Analysis of four degree of freedom Upper Limb Exoskeleton Robot, January 2014, https://www.researchgate.net/publication/259911128.

[89] Bueno, L; Brunetti, F; Frizera, A; and Pons, JL. Human-robot cognitive interaction, In: Pons, JL. *Wearable Robots: Biomechatronic Exoskeletons.* John Wiley and Sons; 2008; 87 - 125.

[90] Xianzhi, J; Caihua, X; Ronglei, S; and Youlun, X. Fuzzy Hybrid Force-Position Control for the Robotic Arm of an Upper Limb Rehabilitation Robot Powered by Pneumatic Muscles, *2010 International Conference on E-Product E-Service and E-Entertainment (ICEEE),* 7th -9th November 2010, Henan: China, 1 – 4.

[91] Rosen, J; Perry, JC; Manning, N; Burns, S; and Hannaford, B. The Human Arm Kinematics and Dynamics during Daily Activities – Toward a 7 DOF Upper Limb Powered Exoskeleton, *12th International Conference on Advanced Robotics*, 18th -20th July 2005, Seattle: WA, 532 – 539.

[92] Riener, R; Nef, T; and Colombo, G. Robot-aided neurorehabilitation of the upper extremities. *Medical and Biological Engineering and Computing,* February 2005, 43(1), 2-10.

[93] Cameira˜o, MS; Badia, SB; Duarte, E; Frisoli, A; and Verschure, PFMJ. The Combined Impact of Virtual Reality Neurorehabilitation and Its Interfaces on Upper Extremity Functional Recovery in Patients with Chronic Stroke. *Stroke*, October 2012; 43(10), 2720-2728.

[94] Byl, N; Abrams, GM; Pitsch, E; Fedulow, I; Kim, H; Simkins, M; Nagarajan, S; and Rosen, J. Chronic stroke survivors achieve comparable outcomes following virtual task specific repetitive training guided by a wearable robotic orthosis (UL-EXO7) and actual task specific repetitive training guided by a physical therapist. *Journal of Hand Therapy*, October-December 2013, 26 (4), 343-352.

[95] Otaka, E; Otaka, Y; Kasuga, S; Nishimoto, A; Yamazaki, K; Kawakami, M; Ushiba, J; and Liu, M. Clinical usefulness and validity of robotic measures of reaching movement in hemi paretic stroke patients. *Journal of NeuroEngineering and Rehabilitation*, 12th August 2015, 1-10.

[96] Hu, J; Lim, YJ; Ding, Y; Paluska, D; Solochek, A; Laffery, D; Bonato, O; and Marchessaul, R. An Advanced Rehabilitation Robotic System for Augmenting Healthcare. *33rd Annual International Conference of the IEEE EMBS*, Boston, Massachusetts: USA, 30th August – 3rd September 2011, 2073-2076.

[97] Carey, JR; Durfee, WK; Bhatt, E; Nagpal, A; Weinstein, SA; Anderson, KM; and Lewis, SM. Comparison of finger tracking versus simple movement training via telerehabilitation to alter hand function and cortical reorganization after stroke. *Neurorehabilitation and Neural Repair*, May-June 2007, 21(3), 216-232.

[98] Scudellari M., Google Glass Gets a Second Life in the ER. 25 May 2016. IEEE Spectrum. URL: http://spectrum.ieee.org/the-human-os/biomedical/devices/google-glass-gets-a-second-life-in-the-er.

[99] Troncossi, M; Mazzotti, C; Foumashi, MM; Zannoli, D; and Castelli, VP. Design and Manufacturing of a Hand-and-Wrist Exoskeleton Prototype for the Rehabilitation of Post-Stroke Patients, In: Meneghetti, U., Maggiore A., Parenti, C., Vincenzo Quaderni del DIEM – GMA. Atti della Sesta Giornata di Studio Ettore Funaioli. Bologna: Asterisco; 2012; 111-120.

INDEX

A

B

C

D

E

M

N

O

P

R

S

T

V

W